AF706104

Drones in Whiteout

Drones in Whiteout

Autonomous Systems Built for the Harshest Theatre on Earth

By

Jonas Merakai

Vij Books
New Delhi (India)

Published by

Vij Books
(Publishers, Distributors & Importers)
4836/24, 3rd Floor, Ansari Road
Delhi – 110 002
Phone: 91-11-43596460
Mobile: 98110 94883
e-mail: contact@vijpublishing.com
www.vijbooks.in

Copyright © 2026

ISBN: 978-93-7794-074-4 (PB)

All rights reserved.

No part of this publication may be reproduced, distributed, or transmitted in any form or by any means — including photocopying, recording, or other electronic or mechanical methods — without the prior written permission of the publisher, except in the case of brief quotations used in reviews or permitted by applicable copyright law.

This book was developed using advanced writing tools to support the drafting and ideation process. All content has been thoughtfully curated, refined, and reviewed to ensure quality and relevance.

Contents

Introduction

On 3 August 1958, USS *Nautilus* passed beneath the North Pole, a cold machine threading under a ceiling of ice that had long stood for remoteness itself. Inside the submarine, the air would have carried the familiar smell of oil, metal and men at work; outside was a world that did not care whether instruments were elegant or crews were brave. The feat became a symbol of technical audacity, but it also revealed an older truth about the polar regions. They reward ingenuity, certainly. They punish dependency more severely. The horizon had simply gone.

That is the paradox at the heart of Arctic autonomy. The farther north the operations push, the more valuable unmanned systems appear. They can fly where crews should not be risked, watch where people cannot linger, and stay on station longer than human bodies comfortably endure. Yet the same environment that makes such systems desirable also strips away the conditions on which they most rely. Cold weakens batteries and stiffens seals. Snow blinds cameras. Ice shifts the meaning of a map. Distance turns a trivial maintenance task into a logistical event. Communications thin out, then falter. Cold edits every design.

The modern appetite for drones in the High North is easy to understand. Shipping lanes, fisheries, subsea infrastructure, scientific observation, search and rescue, sovereignty patrols and military deterrence all press into the same geography, each with its own tempo and political weight. An aircraft that can launch without a crew on board, a surface vessel that can patrol for weeks, an underwater vehicle that can measure beneath ice where no ship can follow: these are not luxuries in the Arctic. They look like common sense. But Arctic common sense has always been stern. It asks a simple question before it permits ambition to proceed: what still works after the margin has gone?

The answer is rarely found in rhetoric about machine intelligence. It is found in the coarse-grained materials and operations. A polymer housing that remains flexible in a temperate workshop may craze after repeated cold soak and vibration. A connector that clicks neatly into place on a bench may not fully seat when a technician is wearing gloves in darkness and blowing snow. A battery that appears healthy on a specification sheet may sag under load during climb-out because its internal resistance has increased with age and temperature. An optical payload may provide beautiful imagery in a demonstration video and nearly useless ambiguity in flat light over sea ice. Trust, in such conditions, is not a philosophical mood. It is the practical consequence of repeatability.

There is a reason seasoned operators tend to speak less grandly about autonomy than brochures do. They learn, often quickly, that the decisive question is not whether a platform can act on its own in a narrow technical sense. It is whether the whole system can remain useful when several small degradations arrive together. The Arctic is exceptionally good at combining them. A little frost on a sensor cover, a little drift in inertial estimation, a little extra wind on recovery, a little less bandwidth than expected, a little moisture where moisture should not be: each problem seems manageable in isolation. In combination, they change the mission. Nothing stays local for long.

Seen clearly, then, Arctic autonomy is less a triumph of disembodied software than a test of engineering discipline under environmental coercion. The setting is austere, but the underlying logic is ordinary and unforgiving. Systems inherit the properties of their parts, and missions inherit the weaknesses of their systems. The result is that platform design choices become operational constraints long before anyone talks about tactics. The true ceiling on delegated control may be set not by what an algorithm can classify, but by how often a gimbal ices, how accurately a vehicle can navigate when visual references thin out, how fast spare seals can be moved to a remote site, or how predictably a lost-link routine behaves when a safe return path no longer resembles the one stored in memory.

There are recent examples that sharpen the point. In 2021, NOAA and Saildrone sent uncrewed surface vehicles into waters off Alaska and the wider Arctic to collect meteorological and oceanographic data over long periods at sea. It was a striking demonstration of what autonomy can do well in the polar domain when endurance, remote sensing and reduced human exposure matter more than rapid manoeuvre. Show the case plainly, and its concept becomes visible: long-duration presence. Name the real achievement, and it becomes even clearer: persistence without a large embarked crew. Explain why it worked, and the answer lies not in some magical immunity to the environment but in a careful fit between task and platform. A surface vehicle does not solve Arctic weather; it chooses a mission that can absorb certain kinds of weather while managing energy, corrosion, routing and data return in a disciplined way. The implication is important. Success in the High North is often less about maximising autonomy than about narrowing the demand placed upon it.

The temptation, especially in policy debate, is to treat autonomy as a scalar quantity, as though more were simply better. The Arctic exposes the poverty of that assumption. An aircraft that can identify leads in sea ice, survey a coastline, or relay a communications link may perform admirably at altitude and still become operationally suspect if launch and recovery remain fragile in gusts, icing, or low contrast. A navigation stack that appears stable across a short test route may accumulate enough error over longer endurance to force conservative geofencing and large safety buffers. A vehicle that saves crew exposure during the sortie may create intense exposure before and after it, when technicians must heat batteries, inspect control surfaces, dry connectors, clean optics, upload routes, check logs, and decide whether a barely reproducible fault is random noise or the beginning of a pattern. Efficiency, here, is not a slogan. It is measured in sorties generated without incident, in mean time between failures, in recoveries completed without drama, and in data usable on return.

A more sobering illustration lies in visibility itself. Whiteout is often described as weather, but operationally, it is closer to a collapse of reference. The eye loses edges; distance estimation goes strange; surface and sky blend into a single indifferent field. Human pilots have long feared it for good reason, yet the machine inherits its own version of the same trouble. Cameras lose contrast. Horizon estimation degrades. Feature-matching struggles when the terrain repeats itself in tone and texture. Low-altitude operations become fraught precisely where many missions become most dependent on them: launch, landing, inspection, close observation. When the horizon is erased, oversight weakens at the very moment when systems need it most. Risk is no longer an abstract noun. It appears as a delayed wave-off, a missed cue in a video feed, an unnecessary second approach in crosswind, an avoidable hard landing that becomes tomorrow's maintenance burden.

History offers a useful corrective to technological self-confidence. The Arctic has always demanded that human plans submit to physical reality. The 1925 serum run to Nome, Alaska, conducted by relays of dog teams across winter conditions, has entered public memory as a story of grit; it also deserves to be read as a story about logistics and reliable movement when infrastructure is sparse and delay kills. Different century, different machines, same governing logic. A system matters because it can be sustained through bad weather, distance and uncertainty, not because it looks impressive at the point of departure. The vehicles have changed. The arithmetic has not.

And yet, it would be shallow to claim that harsh conditions simply defeat autonomy. They do not. A credible counter-example sits far from the Arctic but close to the engineering question. In April 2021, NASA's *Ingenuity* helicopter made the first powered flight on Mars in Jezero Crater, operating with extraordinary autonomy under severe environmental constraints, including cold and communication delay. That achievement matters because it proves that machines can function brilliantly in places hostile to direct human control. Still, the lesson is not that Arctic operations should imitate Martian ones without remainder. *Ingenuity* succeeded within a tightly bounded mission architecture, with exquisite preparation, narrow tasking, and none of the messy operational pluralism that defines the High North: icing, salt spray, moving ships, shifting legal authorities, field repairs with gloves on, contested emissions, and the requirement not merely to demonstrate flight but to generate repeatable, governable service across a season. The limitation is instructive. Harshness alone is not the issue. Coupled dependencies are.

That phrase matters because Arctic systems rarely fail in a single, theatrical way. They fray. A maintenance log accumulates small anomalies. Power margins narrow after repeated cycles. Sensor calibration drifts just enough to remain invisible until GNSS quality worsens. A crew begins to rely on workaround habits that were never designed, only inherited. By the time a platform is declared unreliable, the causes are often dispersed across design, operation and sustainment. One of the most consequential habits in cold-weather programmes is therefore not daring but record-keeping: telemetry that can be trusted, defect reports that are specific, returned-part analysis that is prompt, and configuration control strict enough to distinguish one fault family from another. There is nothing glamorous in that sentence. There is a great deal of operational truth in it.

The Arctic also compresses the moral distance between control and consequence. In gentler settings, organisations can afford to blur who is responsible for what. In severe settings, ambiguity becomes dangerous. If a communication link drops, what authority has already been delegated to the vehicle, and what remains meaningfully under human supervision?

If navigation confidence degrades, should the platform continue toward mission objectives, divert, loiter, descend, climb, or terminate the sortie? If a recovery attempt becomes unstable, what conditions trigger an abort, and who is empowered to make that call when video, telemetry and environmental cues disagree? These are not merely technical branches in software logic. They are governance choices expressed in code and procedure. The machine acts, but an institution has already decided the shape of that action.

That is why autonomy in the High North is, before anything else, an argument about limits. Not limits in the defeatist sense, as though the environment were only a wall, but limits in the engineering sense: measurable boundaries that define where a system remains safe, useful and recoverable. Wind envelopes. Temperature envelopes. Icing envelopes. Communication envelopes. Maintenance envelopes. The most responsible programmes do not treat these as embarrassments to be hidden behind broad claims of capability. They treat them as the conditions of honest performance. Distance magnifies every error.

Once this is seen, several familiar debates look different. Questions about sensors become questions about contamination, calibration and confidence reporting. Questions about batteries become questions about thermal budgets, charging discipline and sortie generation. Questions about navigation become questions about reference scarcity, integrity monitoring and the size of safe corridors when maps, inertial drift and magnetic behaviour disagree. Questions about interoperability become questions about whether data can be stored, tagged, moved and audited across fragile links and different organisational standards. Even cyber hygiene appears in a new light: not an abstract domain apart, but another strand in the effort to keep a system available, legible and trustworthy when the field site is cold, the shelter cramped, and the appetite to postpone a patch forever present.

None of this diminishes the strategic significance of unmanned systems in polar operations. It makes that significance more intelligible. The Arctic is not becoming easier, quieter or less contested. Climate shifts alter access without making the environment hospitable. Thaw in one place may create mobility while freeze, fog and storm still constrict action elsewhere. More actors arrive, more infrastructure matters, more observation is demanded, and the political cost of failure can be disproportionate to the size of the machine that fails. A drone lost to an avoidable recovery misjudgement is not only a line item on a maintenance report. It may be a gap in surveillance, a degraded rescue capability, a diplomatic embarrassment, an intelligence loss, or a public demonstration that a system advertised as reliable was merely lucky in rehearsal.

The harder, better question is therefore not whether autonomous systems belong in the Arctic. They already do. The question is what sort of autonomy deserves to remain there after the demonstrations, after the procurement slides, after the first winter season exposes all the minor assumptions that temperate testing left untouched. Some answers lie in hardware: choosing parts that survive derating, sealing, icing, vibration and glove-handling with dignity. Some lie in software: designing graceful degradation rather than theatrical precision that vanishes in bad weather. Some lie in operations: structuring routes, geofences, crew roles and lost-link behaviour so that surprises do not cascade. Some lie in institutions: accepting that maintenance pacing, spares forecasting, data custody and field feedback shape operational reality as powerfully as sensor sophistication does. The erased horizon returns here, quietly, as a diagnostic image. When reference thins out, whatever remains dependable is what the mission truly rests upon.

There is a stern kind of optimism inside that diagnosis. It refuses fantasy, but it does not refuse progress. Machines can become more dependable in the cold. Testing can become more honest. Data from failures can be organised into learning rather than blame. Programmes can be designed across seasons rather than around one triumphant deployment. Control can be allocated with greater humility and greater precision. The Arctic does not ask for perfection. It asks for candour about conditions, discipline about limits, and respect for the slow labour by which reliability is earned.

That labour is where serious judgement begins. Faced with a platform advertised as Arctic-ready, the meaningful inquiry is not whether it can fly, sail or sense under idealised conditions; many systems can do that. It is whether it can be maintained, recovered, trusted and governed when visibility collapses, power thins, links break, parts age, and people must decide under pressure what may still be delegated to the machine. The trade-off is rarely between ambition and caution. More often it is between spectacle and service. On an erased horizon, that difference is everything.

Chapter 1

Arctic Operating Environment Constraints

What does it mean to call a system autonomous in a place that steadily strips away the conditions autonomy prefers? The question sounds technical until one imagines the ordinary sequence on which confidence depends: a battery delivers what its indicator promised, a seal remains supple, a lens stays clear, a connector mates cleanly, a link holds, a landing surface looks like a landing surface. In the Arctic, each of those assumptions comes under pressure at once. A machine may still move, sense, and calculate, yet the ground beneath its claims has already shifted. The puzzle is not whether a vehicle can act without a hand on the controls. The puzzle is whether the world around it still permits those actions to mean what designers thought they meant.

That pressure rarely arrives with theatrical clarity. More often, it gathers in quiet, stubborn ways. Cold changes the feel of materials and the timing of mechanisms. Wind turns a manageable procedure into a narrow gamble. Snow enters places that looked closed in the workshop. Visibility thins until judgement itself becomes harder to anchor. A small fault that would be tedious elsewhere becomes serious here because distance lengthens every recovery, every inspection, every replacement, every second attempt. The Arctic does not merely add difficulty. It rearranges it.

The sharpest misunderstandings about unmanned systems in severe environments begin with the belief that tactics sit at the centre and engineering follows behind. In easier settings, that illusion can survive for a while. A capable crew can compensate, improvisation can bridge small gaps, and maintenance can absorb the cost later. In the High North, the order is reversed. Physical behaviour sets the terms long before doctrine declares intent. Plastics, lubricants, fasteners, housings, bearings, intake paths, antenna lines, batteries, and seals each carry their own verdict on what is feasible, and they do so without regard for mission plans. The environment is not background scenery. It is an active editor.

That is why the stakes in this terrain are larger than simple platform performance. When a system behaves unpredictably in cold and low contrast, the problem does not remain trapped inside a single component. It moves outward. The operator hesitates over whether an anomaly is real. The maintainer inherits faults that resist reproduction. The planner loses confidence in timelines that once looked sensible. Spare parts suddenly matter more than payload brochures. A delayed launch, a waved-off recovery, a diverted route, a shortened sortie, an unexpected inspection - each one may look minor in isolation, but together they decide whether presence can be sustained. Reliability, in such conditions, is not a narrow engineering virtue. It is the basis of operational truth.

The matter grows more serious because the Arctic is unforgiving of hidden dependencies. Many systems are praised for independence when what they actually possess is a chain of support that remains invisible in benign conditions: stable power, clean launch surfaces, predictable access to shelter, uninterrupted data paths, easy handling, rapid repair, and familiar visual cues. Remove those supports, and a supposedly self-sufficient machine begins to reveal how much assistance it quietly expected all along. A drone that flies well but cannot be turned around quickly in blowing snow, a sensor that detects well but loses alignment after thermal cycling, a control station that functions neatly indoors but becomes clumsy under gloves and fogged screens - these are not separate disappointments. They belong to one argument about what autonomy really costs when the weather takes away the margin.

Still, harshness alone does not make Arctic operations mysterious. Much of the difficulty becomes clearer, not murkier, once one stops searching for a single master cause. Failure here is usually cumulative. A little contamination, a little stiffness, a little drift, a little uncertainty in the picture, a little extra time spent outside the shelter: the system absorbs each burden until it cannot. The dangerous temptation is to treat these as annoyances rather than a structure. But structure is exactly what they are. They shape how machines age, how crews decide, how often procedures can be repeated, and how far confidence can travel from one successful mission to the next.

The path through this chapter follows that structure as it appears in the field rather than as it is often advertised. It begins at the level of matter itself, where cold, wind, and thermal cycling alter behaviour before any operator can compensate. It then moves into icing, blowing snow, and contamination, where the environment enters the machine, blurring the line between mechanical fault and system instability. From there, the focus tightens on visibility collapse, where whiteout and near-whiteout disturb not only seeing but timing, judgement, and the pace at which operations remain safe. The movement ends with distance and sparse infrastructure, where every weakness acquires a logistical shadow and every recovery option narrows.

A useful way to hold these problems together is to think of a glove. It is an everyday object, unremarkable until conditions turn hostile. A glove protects the hand, yet it also dulls touch, slows fine work, changes grip, and exposes which tasks were designed with real conditions in mind and which merely assumed convenience. The Arctic acts in much the same way on unmanned systems. It does not simply threaten them from the outside; it reveals which claims depended on delicacy, close support, and forgiving circumstances. What seemed precise may prove fussy. What seemed minor may prove decisive.

From that perspective, the environment does more than challenge machines. It disciplines the people and institutions that place faith in them. Every design choice acquires an operational accent. Every maintenance shortcut writes a future cost. Every assumption about launch, recovery, sensing, routing, or repair becomes a wager against cold, distance, and degraded perception. That wager may be worth making. But it ought to be recognised for what it is.

The question that follows is not whether the Arctic is hard. That is too obvious to be interesting. The real question is narrower, harsher, and more useful: when cold begins to stiffen, snow begins to intrude, and the visible world starts to flatten into white, which part of the system fails first - the material, the sensor, the procedure, or the confidence that held them together?

Cold, wind, and materials behaviour

In February 2021, NASA's *Perseverance* rover landed in Jezero Crater carrying a small experiment that would later become famous: the *Ingenuity* helicopter. Mars is not the Arctic, but the engineering lesson is unexpectedly close to it. A flying machine that is asked to survive extreme cold must first survive itself. Batteries cool when left idle. Materials contract at different rates. Lubricants thicken. Tiny clearances stop being tiny. Before a rotor turns or a line of code makes a decision, matter has already voted. Physics speaks first.

That sequence matters in the High North because the seduction of autonomy encourages a dangerous order of attention. People notice the camera feed, the route plan, the machine's apparent independence, and the relief of keeping a crew out of risk. They notice behaviour. They do not always notice what permits behaviour to remain consistent from one sortie to the next: the pliancy of a seal, the grip of a fastener after repeated thermal cycling, the drag in a bearing, the way a cable jacket hardens, the way a housing shrinks around an interface it no longer quite matches. Small shifts. Big consequences.

The Arctic makes design assumptions audible. A connector that felt definite in the workshop begins to feel vague through gloves. A latch that snapped shut in temperate testing now resists, then suddenly over-travels. A polymer cover that looked inert takes on a dry, chalky brittleness. A mechanism that moved with a low, confident hum develops hesitation, then a faint stick-slip chatter. The sound is often the first warning: not failure, not yet, but friction where motion used to be easy. That is why reliability in severe cold begins with unglamorous questions. What contracts? What stiffens? What embrittles? What stops sliding when the temperature falls and the wind keeps stripping away such warmth as the system can retain?

Low temperature changes materials not by magic but by energy. Molecules move less. Polymers lose flexibility as they approach or pass through transitions in their mechanical behaviour. Elastomeric seals that once compressed and rebounded can harden enough to lose intimate contact with mating surfaces. Adhesives may keep their bond in one stress state and lose it in another. Metals contract too, though less dramatically than many plastics, and that difference matters because systems are assembled from mixtures, not pure categories. An enclosure may be aluminium, its cover polymer, its seal elastomer, its fasteners stainless steel, its internal supports some glass-filled composite chosen for lightness and cost. Each part responds to cold according to its own properties. The machine does not cool as one body. It cools as an argument among materials.

That argument often appears first as shrinkage, though shrinkage is not itself the problem. Every competent engineer expects dimensions to change with temperature. Trouble begins when dimensional change becomes differential and repeated. A plastic housing can contract more than the metal insert it surrounds. A gasket groove can alter just enough to reduce sealing pressure. A cable boot can pull away microscopically from a connector body, not enough to advertise a gap, enough to admit moisture over time. Once thermal cycling enters the picture, the matter becomes more serious. Arctic systems do not simply become cold and remain cold. They warm in transport, cool on exposure, warm in shelters, cool again in operation, then endure the local gradients imposed by internal electronics, battery heaters, sunlight, wind and snow loading. A system living through those cycles is not just experiencing temperature. It is enduring continual rearrangement of fit.

The effect on plastics is especially revealing because plastics were supposed to simplify life. They cut weight, resist certain forms of corrosion, accept intricate shapes, and often lower manufacturing costs. Yet the Arctic shows how sharply polymer choice governs reliability. Housing that shrugs off routine handling at mild temperatures may crack at a boss, hinge, clip, or corner under loads that once felt trivial. The failure may appear capricious because the part survived many prior uses. It was never capricious. Temperature narrowed the margin until ordinary handling crossed it. One of the quiet disciplines of cold-weather engineering lies in refusing to ask brittle parts to perform tactile, forgiving jobs. Snap fits that seem elegant in a design review can become liabilities in a field shelter at the end of a long shift. Thin tabs punish cold. So do decorative geometries mistaken for structure.

Seals deserve equal suspicion because they sit at the boundary between a machine's clean interior and the Arctic's persistent contamination. A seal is a promise made in compression. It works because material softness, geometry and surface condition align closely enough to resist ingress. Cold interferes with each of those terms. Elastomers stiffen. Surface irregularities matter more. The compression set becomes more consequential if the seal no longer springs back cleanly after repeated openings. The result is not always a dramatic leak. More often it is slow betrayal: fine moisture entering during handling, then freezing; condensation forming where it should not; snow melt reaching a connector recess; a pressure difference drawing contamination through a path too small to notice in inspection. The point is not that seals fail in the cold. The point is that the meaning of "sealed" becomes narrower and more conditional than bench tests imply.

Lubricants tell the same story in a different language. In ordinary conversation, lubrication sounds simple: reduce friction, protect surfaces, keep parts moving. In severe cold, the properties that matter become starkly time-dependent. Viscosity rises. Flow changes. Start-up torque increases. A motor that has enough authority once warmed may struggle during the first moments of motion, precisely when the control system assumes responsiveness. Bearings that seemed free-running can show drag. Gear trains may not seize outright; they may hesitate, then release, then hesitate again. That behaviour has a name engineers know well: stiction, the stubborn friction of things reluctant to begin moving. In an Arctic drone, stiction turns up in places operators experience as symptoms rather than causes. A gimbal overshoots because it did not start smoothly. A control surface lags by a fraction that matters during gust response. A fold-out mechanism deploys unevenly. By the time software notices a discrepancy, the source may be buried in chemistry and surface contact, not code.

Wind complicates this physics in a way that public discussion often blurs. Wind chill is a phrase built around human sensation. It does not cause inanimate objects to fall below ambient air temperature merely by blowing harder. Yet wind matters enormously because it accelerates heat loss, erases local warm pockets, intensifies temperature gradients, and changes how quickly a system reaches equilibrium with its surroundings. A battery pack sheltered inside a fuselage may hold useful warmth when still air allows internal heat to linger. Expose the same airframe to sustained cold wind, and the thermal budget shifts. Surfaces cool faster. Seals stiffen sooner. Lubricant viscosity stays elevated for longer. Condensation patterns can change when a vehicle moves from shelter to open air or from powered operation to shutdown. A technician may experience wind as discomfort; the machine experiences it as a more aggressive timetable.

That timetable helps explain why cold-weather failures often cluster around transitions rather than steady-state operation. Launch after exposure is dangerous because components have not yet found a stable thermal condition. Recovery is dangerous because warmed internal spaces may meet cold moisture-laden air at precisely the wrong moment. Maintenance is dangerous because disassembly and reassembly disturb fragile interfaces while parts sit near the edge of their intended mechanical behaviour. Even storage can be treacherous if repeated freeze-thaw cycles and incomplete drying accumulate invisible damage. The familiar wish that a system either work or fail cleanly does not survive here. Arctic degradation prefers ambiguity.

A revealing example arrived from a different cold theatre but with lessons that travel north. During the 1986 Space Shuttle *Challenger* accident, attention later centred on the solid rocket booster O-rings and the effect of unusually low temperature at launch on seal performance at Kennedy Space Centre in Florida. The context was not Arctic, the system was vastly more complex, and the stakes incomparable in scale. Yet the mechanical warning remains brutally relevant. Elastomeric behaviour that appears acceptable across a wide operating envelope can become fatally slow or incomplete when temperature narrows resilience and timing. That episode does not prove that every cold-weather seal problem leads to catastrophe. It proves something more useful: materials do not respect organisational confidence. When low-temperature changes in response, managerial optimism cannot close the gap.

A second example, closer in spirit to fielded autonomous systems, emerged repeatedly in polar and high-latitude operations involving small unmanned aircraft used by research teams and public agencies. The pattern is familiar to anyone who has read technical notes from cold-weather deployments: airframes may survive the cold better than the supposedly minor subsystems mounted on them. Camera housings fog. Servo response slows. Cable insulation stiffens enough to alter strain relief and connector reliability. None of these problems necessarily destroys the platform. They do something more corrosive to operations. They make performance intermittent. The machine becomes difficult to trust not because it always fails, but because it stops failing in a consistent way. The maintainer cannot easily reproduce the defect indoors. The operator cannot easily predict when a slight control irregularity is harmless and when it signals the onset of a broader loss of margin. Intermittence is poison in austere operations because it consumes decision-making time.

It is tempting, at this point, to imagine a counter-example strong enough to cut through the severity of the argument. After all, plenty of machines work in cold regions. Ski bindings flex. Snowmobiles run. Aircraft operate in northern winters. Surely materials science and good practice already solve much of this. That objection matters because it keeps the discussion honest. Severe cold does not make reliable operation impossible, nor does every polymer turn brittle or every lubricant turn treacherous at the same threshold. Careful material selection, proper derating, suitable low-temperature greases, thoughtful tolerance stacks, disciplined pre-heating and realistic testing can produce systems that perform impressively. Some designs cope very well indeed.

But the counter-example only holds if one notices its hidden condition: those systems work because someone made cold behaviour central to the design brief rather than treating it as a late qualification nuisance. That is the reconciliation. Reliability in the Arctic is not blocked by physics; it is organised by physics. The environment is not announcing impossibility. It is demanding specificity. Which polymer grade? Which seal compound? Which grease, in which bearing, at what start-up temperature, after how many cycles? Which fastener preload remains trustworthy after contraction and vibration? Which interface still closes properly when handled outdoors with reduced dexterity? Once the questions become this concrete, the notion of a generic "cold-capable" system begins to look thin.

Fasteners reveal the point with almost cruel clarity. They are mundane, cheap, and omnipresent. They also hold together a great deal of the confidence people place in machines. Threaded joints depend on preload, friction, material pairing, surface condition and assembly technique. Cold alters those relationships. Differential contraction can change clamp force. Repeated cycling can loosen margins originally judged acceptable. A fastener installed correctly in mild conditions may behave differently after vibration and cold exposure, particularly where polymers creep, inserts shift, or mating parts settle. Glove-handling adds another layer: torque feel worsens, dropped hardware becomes more likely, thread starts are easier to miss, and field rework grows less repeatable. When operators speak darkly of "mysterious looseness" in cold-deployed equipment, the mystery usually lies not in the fastener but in the assumptions surrounding it.

The broader lesson is almost austere in its simplicity. Tactics cannot rescue a machine from a bad coefficient of expansion. Autonomy cannot negotiate with thickened grease. Clever interfaces do not repeal embrittlement. A programme may speak grandly of persistence, sensing and delegated control, but if the material stack underneath those ambitions was chosen for temperate convenience, the Arctic will eventually expose the bargain. It does so not with theatrical vengeance but with attrition: a crack here, a delayed motion there, an ingress path no one noticed, a mounting point that loosens half a turn at a time, a battery compartment that never quite recovers the warmth it assumed. The sound returns - that faint change in how a mechanism moves, the small warning before the obvious fault.

That is why the real test begins so early. Not when the aircraft reaches altitude, not when the route looks elegant on a display, not when operators praise the video feed, but when cold first edits fit, motion and contact. The serious question is never simply whether a system functions at low temperature. It is whether its materials still behave with enough consistency that operators can commit to it again tomorrow, after another cycle of cold, wind, handling and waiting. If the answer depends on luck, the design is already speaking too softly to the surrounding climate.

Icing, blowing snow, and contamination

On 13 January 1982, Air Florida Flight 90 left Washington National Airport with ice still compromising the aircraft, and within moments it crashed into the 14th Street Bridge over the Potomac. The event belongs to civil aviation history, not Arctic drone operations, yet its lesson travels north with undiminished force. Ice does not need to look dramatic to become decisive. A thin layer changes shape, weight, flow and response. It alters what a wing means. It alters what an engine can ingest. It alters what a crew or an operator thinks they are controlling. Contamination rarely arrives alone.

That last point matters more in the High North than public discussion often admits. When people picture cold-weather failure, they imagine a clean contest between machine and temperature: batteries losing strength, plastics hardening, lubricants thickening. Those effects are real, but snow and ice add a crueller difficulty. They do not merely cool the system; they enter it, adhere to it, melt, refreeze, drift, obscure, abrade and migrate. Fine snow behaves less like weather than like a patient contaminant. It follows airflow into intakes. It settles around seals. It packs into corners that inspection misses. It clings where warmth briefly turns crystals tacky, then hardens again when the temperature drops. A machine may remain mechanically alive while its margins quietly shrink.

That is why icing deserves to be understood not as a seasonal inconvenience but as an aerodynamic and systems problem with many entry points. Show the process plainly and the concept becomes difficult to ignore. A wing, rotor blade or propeller acquires roughness and altered contour; lift falls, drag rises, vibration can increase, and control response changes. Name the mechanism and it becomes more exact: accretion. Explain it and the picture sharpens further. Supercooled water droplets, wet snow, freezing drizzle, sea spray in cold air, and repeated transitions through moist layers can all deposit material where the design assumed clean flow. The implication is not merely that performance worsens. It is that the aircraft begins to depart from its tested self. Numbers used in planning no longer describe the object in the sky.

For large crewed aircraft, this truth has long been written in procedure, training and accident investigation. For smaller unmanned systems, it can be easier to underestimate because the machines are lighter, simpler and often presented as expendable enough to accept risk. Yet lightness offers no exemption from icing physics. A small rotorcraft or fixed-wing drone may, in fact, have less surplus power, less thermal capacity, and less tolerance for changes in mass distribution or surface roughness than a larger platform. A modest ridge of ice on a propeller blade can produce an imbalance. A little contamination on a pitot opening or pressure port can distort the air data on which flight logic depends. A camera dome with frozen residue may still transmit an image, but not one an operator should trust. The visible symptom comes late.

Blowing snow deepens the problem because it defeats the comfortable distinction between exterior weather and interior machinery. Snow in the Arctic is not always the soft, picturesque blanket imagined by outsiders. It can move laterally at speed, driven by wind into a fine abrasive cloud that finds seams, hinges and vents with astonishing persistence. The scene and its logic are obvious: a vehicle taxis, idles, lands, or sits briefly on an exposed surface while dry snow streams along the ground and swirls up into every low-pressure pocket. Name the process, and it becomes engineering rather than scenery: ingress. Explain the mechanism, and one sees why design claims fail. Enclosures breathe as temperatures shift. Access panels never seal perfectly forever. Cable runs and rotating assemblies create pathways. The implication is serious because ingress rarely announces itself during the mission in which it begins. It declares itself later, after melt, refreeze, corrosion, drag or intermittent electrical behaviour has had time to mature.

A camera gimbal offers a particularly revealing example because it sits at the intersection of sensing, control and operator judgement. In benign conditions, a gimbal is easy to describe: it stabilises the image, keeps the line of sight where it is commanded, and supports the human impression that the vehicle remains composed. Add fine snow, however, and the matter changes. Snow packs into bearings or along exposed interfaces, melts slightly from motor warmth or sunlight, then refreezes or turns to slush and grit. The gimbal begins to drift, hesitate, or over-correct. At first, the problem looks optical. The picture wobbles or drifts; the operator suspects software, turbulence, or perhaps a calibration quirk. But the defect is mechanical contamination, and because the gimbal feeds the operator's understanding of vehicle attitude and scene stability, a sensor problem becomes a control problem. Trust turns tangible here. It appears as a delayed decision, a cautious recheck, an unnecessary second pass, a reluctance to commit to what the image appears to show.

The wider machinery is vulnerable in similar ways. Intakes do not need to clog completely to become dangerous. A partial blockage can alter cooling, reduce performance, or change the operating environment of internal components enough to provoke further faults. Electric motors may not ingest snow in the way combustion systems do, but they still suffer when contamination reaches bearings, windings, connectors or cooling paths. Connectors are especially treacherous because they turn contamination into intermittence. A little moisture, a little packed snow, a little corrosion at a contact interface: the result may not be a clean failure but a signal path that degrades only under vibration, only at certain temperatures, only after handling. Those are the hardest faults to diagnose in the field. They refuse to remain still for inspection.

One reason intermittent faults are so corrosive to operations is that they damage organisational memory. A part that fails cleanly teaches. A part that behaves badly only after a cold landing in blowing snow, then appears healthy indoors, teaches very little unless maintenance discipline is unusually good. The operator reports a transient warning. The maintainer checks the system in the shelter and finds nothing decisive. The next sortie goes ahead because the evidence feels too thin to justify grounding. Then a similar irregularity appears in a different place, or not at all, and the programme begins to rely on anecdote. This is where contamination exerts one of its subtlest effects. It does not merely break equipment. It weakens confidence in diagnosis.

The history of aviation offers another hard anchor. On 10 March 1989, Air Ontario Flight 1363 crashed after take-off from Dryden, Ontario, with ice and snow contamination on the wings. The immediate context again involved a crewed aircraft, not an autonomous platform, but the mechanical truth remains blunt. Contaminated surfaces alter lift and control at precisely the phase of flight where margin is already limited. The relevance to Arctic unmanned systems is not rhetorical. Launch and recovery are often the least forgiving parts of the mission for smaller aircraft. They happen low, close to obstacles, with little time to recognise that an apparently minor contamination issue has changed the aerodynamic balance. A drone that cruises capably once established may still be an unreliable operational tool if it reaches that safe state only inconsistently.

This should not lead to fatalism. A fair account has to admit that some systems cope well with icing and contamination, especially when the mission, platform and environment are matched honestly. Larger aircraft have long used thermal anti-icing, pneumatic boots, careful air data design, drainage paths and inspection routines to manage predictable forms of accretion. Some unmanned platforms also incorporate heated sensors, protected intakes, sealed actuators or sacrificial covers that materially improve survivability. Research aircraft and specialised cold-weather systems can perform impressively when designed with contamination as a central requirement rather than an afterthought. That is the necessary counterweight.

And yet the counter-example only strengthens the deeper argument. Where contamination is managed successfully, it is managed through explicit design trade-offs, not optimism. Heating consumes power. Sealing complicates maintenance. Drainage paths compete with ingress paths. Covers and fairings add weight, drag, cost, and sometimes their own failure modes. A cleaner airframe may become a less serviceable one. A tightly sealed connector may become harder to inspect and harder to mate correctly with gloves. Protection never arrives free. The Arctic forces each choice into view because the price of concealment rises so quickly once weather, distance and repeated handling interact.

The same tension governs blowing snow on the ground. It is tempting to treat ground time as safe time, but contamination often accumulates while the vehicle is parked, prepared or recovered. Rotor wash and propeller slipstreams can recirculate loose snow into sensitive areas during start-up or shutdown. Recovery teams rushing to secure an aircraft in poor visibility may brush, lift, fold and pack components while snow is still drifting through open interfaces. A system lands as one machine and enters storage as another, carrying hidden moisture into a warmer enclosure where melt begins. The visible mission has ended; the failure process has not. Fine white dust returns here as a quiet motif of damage: not one storm, but a residue of many brief exposures, each too small to command attention, together enough to alter the machine's future.

Contamination also scrambles the neat division between environmental hardship and adversarial risk. A vehicle already struggling with obscured optics, degraded connector integrity, or uncertain sensor alignment becomes easier to misread and easier to lose without any dramatic external interference. Operators faced with noisy data may interpret a true warning as nuisance or dismiss a nuisance as true warning. Maintenance teams under pressure may swap parts that are innocent while leaving contamination pathways untouched. Planning staff may blame sortie failure on weather in the broad sense when the actionable cause was an avoidable design exposure at a vent, bearing or access point. Precision in language matters because it guides corrective action. "Bad conditions" is often not precise enough. The question is where the snow went.

That question cuts into the politics of trust inside organisations. People often speak of trust in machines as though it were a psychological stance, either excessive or insufficient. In practice it is more physical. Trust appears when an operator keeps the planned route because the airframe has behaved consistently through earlier cold recoveries. It appears when a maintainer signs off a connector after inspection because the sealing scheme has earned respect across a season. It disappears when a gimbal wanders after a storm and no one can tell whether the root cause lies in software, contamination or handling. It disappears when a vehicle returns with one more unexplained anomaly and the team quietly shortens the next mission. Trust has fingerprints, logs, waved-off launches, extra inspections, swapped components and cautious language in the debrief.

Contamination resists the modern taste for clean categories. Is an iced lens a sensor fault or an environmental condition? Is packed snow in a hinge a maintenance lapse or an unavoidable field penalty? Is intermittent motor behaviour a hardware issue, an enclosure issue or a recovery-procedure issue? The honest answer is often all three. Arctic operations punish institutions that allocate blame too quickly because blame flattens mechanism. Better questions stay close to the material sequence. What adhered? What entered? What melted? What refroze? What shifted only after vibration? What changed between exposure and inspection? The more faithfully those sequences are reconstructed, the less mysterious "Arctic unreliability" becomes.

The unresolved difficulty is that the most effective defences against contamination can make a system heavier, less accessible, more power-hungry, and sometimes more complex to maintain in the field. The cleanest sensor window may require heating. The best-protected bearing may be harder to service. The most tightly sealed connector may slow urgent turnaround. So the essential test is not whether contamination can be wished away, but whether a programme knows exactly which surfaces must remain clean, which cavities must stay dry, and which deviations from that ideal it can tolerate without crossing from inconvenience into ungovernable doubt. When snow moves like dust and ice changes shape in silence, that line becomes the one worth measuring.

Visibility collapse and operational tempo

On 28 November 1979, Air New Zealand Flight 901 flew into the slopes of Mount Erebus on Ross Island, Antarctica, killing all aboard. The catastrophe had several causes, but one of the most haunting features of the disaster was visual: a world that could look navigable while withholding the clues a pilot expects. In polar light, snow and cloud can merge into a deceptive field in which distance, slope and horizon stop behaving normally. The mountain does not announce itself. The eye keeps searching for an edge.

That is the beginning of whiteout as an operational condition rather than a poetic description of bad weather. Whiteout is not simply "poor visibility", still less a dramatic synonym for a snowstorm. It is a collapse of usable reference. Contrast drains away. Shadows disappear. Surface and sky lose their boundary. Objects remain present yet become hard to judge in relation to one another. A person can still see, in the strict sense, and still be functionally blind to what matters: angle, closure rate, drift, height above ground, the texture of terrain, the true shape of an approach path. The world has not vanished. Its grammar has.

Near-whiteout can be even more treacherous because it tempts action. Full whiteout often stops operations outright. Near-whiteout does something subtler. It preserves enough visibility to encourage continuation while quietly removing the cues that make continuation safe. A launch still looks possible. A landing zone still seems identifiable. A slope still appears level enough from one angle. A vehicle's video feed still returns an image. Yet the margins have already altered. The machine and the people supervising it begin to act on a scene that resembles clarity more than it deserves. False confidence forms quickly.

For unmanned systems, that distinction matters because both halves of the human-machine relationship lose reference at once. Human oversight weakens when operators can no longer cross-check instruments against the outside world or read a camera feed against a stable horizon. Machine perception weakens because many sensors rely, directly or indirectly, on features, edges, contrast gradients and environmental regularity. Show a low-altitude drone approaching a snow-covered surface in flat light and the concept becomes concrete: the onboard camera sees a pale field with little texture; the operator sees roughly the same; the aircraft's control system may hold attitude perfectly while lacking the richer contextual cues that would reveal whether the descent remains prudent. Name the condition and it sharpens: reference scarcity. Explain it and the operational implication becomes plain. When references thin out, stable flight and safe flight begin to diverge.

This is why launch and recovery deserve more suspicion than cruise. A drone established at altitude with healthy navigation, adequate energy and decent separation from obstacles may tolerate low contrast far better than the same aircraft can tolerate it a few metres above a featureless surface in gusting crosswind. Most missions spend relatively little time in those transition zones, which makes them easy to underweight in planning. Yet that brief fraction often determines whether the sortie counts as success, abort, incident or loss. Tempo suffers here first. Not in the elegant middle of the mission, but at the seams.

Operational tempo is too often discussed as if it were a matter of will, staffing or tasking discipline. In severe environments, it is more truthful to think of tempo as the speed at which uncertainty can be resolved without cheating. A plan that assumes a certain number of daily launches may appear efficient under clear skies, with crisp horizons, predictable recoveries and generous turnaround windows. Add repeated episodes of low contrast and the arithmetic changes. Crews pause longer before committing to launch. Recovery approaches multiply. Divergence between pilot, payload operator and supervisor takes longer to reconcile. Post-flight inspection becomes more exacting because poor visibility during recovery raises suspicion about hard-to-see contact, icing, ingestion or minor structural shock. Nothing dramatic may happen. The schedule still slips.

The 1979 Erebus accident remains important because it shows how visual confidence can survive after visual truth has gone missing. In polar environments, terrain can blend into atmospheric brightness so thoroughly that a large obstacle presents weakly or deceptively. For manned aviation, that insight shaped later caution around flat-light and whiteout operations. For autonomous systems, it should do the same. A machine equipped with stabilised imaging, inertial sensing, GNSS and terrain databases may seem protected against the frailties of the human eye. Sometimes it is. But the protection is conditional. Databases may be coarse. Altitude estimates may not tell the whole story over uneven snow or sea ice. Sensor fusion may inherit common-mode weakness if multiple inputs all degrade in low-contrast, low-feature conditions. A clean display can conceal a fragile understanding of the scene.

A second example comes from helicopter operations in snow-covered terrain, where whiteout during landing has long been recognised as a recurrent hazard. The specific incidents vary by service and country, but the operational pattern is stable enough to be a matter of doctrine in cold-weather flying. Rotor wash lifts loose snow into a dense cloud; visual references disappear; pilots lose the cues needed to judge descent rate, lateral drift and touchdown attitude. The aircraft may be mechanically sound, the crew experienced, the landing zone nominally prepared. Then the horizon goes, replaced by swirling brightness and uncertainty. For autonomous or semi-autonomous systems, the analogue is immediate. Propeller wash, rotor wash, exhaust flow or even the disturbance caused by touchdown can turn a marginally visible scene into an unreadable one at the very phase when control demands sharpen. The system does not merely enter bad weather. It creates some of the perceptual problem itself.

Once that mechanism is recognised, low-altitude flight over snow and ice looks different. It is not only that onboard cameras may struggle to detect texture or landmarks. It is that the physical environment may offer fewer landmarks to begin with, and the few that remain may repeat. Snow-covered ridges, pressure features, ice leads, drifts and flat expanses can produce scenes in which pattern recognition becomes unreliable or overconfident. A person looking at the display may feel that the vehicle is progressing smoothly over a legible surface when in fact scale and distance cues have become thin. The danger is rarely melodramatic disorientation in the cinematic sense. More often it is a small misjudgement preserved long enough to matter: an approach begun too soon, a descent continued too low, an obstacle noticed too late, a recovery lane misread by just enough to require an unstable correction.

The horizon returns here as a quiet organising fact. When the horizon holds, many kinds of error remain containable. Human operators notice attitude more quickly. Video becomes easier to interpret. Even rough visual navigation benefits from a stable sense of "up", "away" and "level". When the horizon dissolves, systems must substitute other forms of certainty. They may use inertial estimates, radar or lidar altitude cues, structured routes, predefined recovery boxes, stricter minima, or conservative go-around logic. All of these can help. None of them abolishes the problem, because visibility collapse is not merely informational. It changes timing, trust and workload simultaneously.

That simultaneity explains why low contrast so often produces automation surprise. Operators supervising an autonomous or highly assisted vehicle tend to relax into the system's rhythm when behaviour remains smooth. Then the environment removes the very cues that would help them anticipate a mode change, recovery trigger or degraded-performance response. A vehicle may slow, hold, divert, climb, reject a landing or insist on a go-around according to perfectly rational internal logic. Yet if the operator cannot see clearly why the machine has done so, the action feels abrupt or opaque. The human response is not abstract "loss of trust". It becomes visible as a hand reaching for manual override, a delayed consent to an automated decision, a repeated query to confirm position, a cautious extension of separation from terrain, or a blanket tightening of weather criteria after one unsettling sortie. That is how tempo drops: not because the machine stopped working, but because explanation lagged behind behaviour.

A fair account must admit a limitation. Visibility collapse is not an absolute defeat for modern systems. Instrument flight, synthetic vision, terrain awareness tools, radar altimeters, disciplined approach design and carefully structured ground control can all reduce dependence on human eyesight. Certain missions, especially those flown at altitude with conservative recovery arrangements, may remain entirely workable in conditions that would be unacceptable for visually intensive low-level tasks. Some platforms are markedly better than others at preserving utility when contrast fades. That matters, and it prevents the argument from curdling into mysticism about polar weather.

But the limitation does not cancel the main point. It sharpens it. Systems can cope with visibility collapse only when designers and operators acknowledge which parts of the mission still depend on the world remaining visually interpretable and which parts can be transferred to instruments with acceptable residual risk. The honest boundary usually lies at the transitions: launch, low-altitude route adjustments, surface inspection, final approach, touchdown, deck landing, confined recovery areas, improvised sites, or any phase in which the vehicle must judge not just where it is in abstract coordinates but how it sits in relation to texture, drift, obstacle and closure. Where that judgement remains fragile, tempo plans based on clear-air assumptions will break first.

Those break points reveal themselves in mundane decisions rather than dramatic failures. A crew delays launch to wait for a little more contrast, then shortens the route to preserve recovery margin. A supervisor rejects an otherwise acceptable landing because the camera feed looks washed out and depth cues feel unreliable. An operator keeps a vehicle higher than planned over a featureless surface, sacrificing image quality for navigational comfort. A maintenance team extends inspection after a hard-to-interpret recovery because poor visual reference may have masked an incipient strike or ingestion event. Efficiency, in this setting, is not the number of flights promised on paper. It is the number safely completed without forcing crews to pretend they knew more than they did.

Near-whiteout punishes pretence especially hard because it makes compromise look temporary. Teams tell themselves they will reassess after launch, after the first leg, after the next pass, after one more approach. Yet each continuation step consumes cognitive and operational reserves. Energy declines. Daylight shifts. Weather moves. Attention narrows. The vehicle may still be inside nominal limits while the operation as a whole has entered a poorer bargain. The visible world has become expensive to interpret. That cost must be paid somewhere: in delay, in reduced ambition, in more conservative routing, in extra separation, in more frequent aborts, in slower recoveries, in broader buffers around obstacles and terrain. If no one pays it explicitly, the bill arrives later as surprise.

It is worth noticing how often language disguises the issue. Organisations speak of "weather delay" as though all meteorological hindrance were interchangeable. It is not. Strong wind, icing risk, precipitation intensity and visibility collapse make very different demands on aircraft, sensors and crews. Whiteout and near-whiteout are peculiar because they attack judgement directly. They erode not only what the system can see, but how quickly people can verify what they think they see. That is why they spread so effectively through a sortie schedule. A machine delayed by low contrast is not merely delayed by nature. It is delayed by the time required to recover confidence honestly.

The strategic temptation is to answer all this with more sensing, more autonomy, more software. Sometimes that is the right direction. Better altimetry, improved perception in degraded visual environments, more explicit confidence reporting, clearer mode logic and stronger route constraints can materially improve safety and usefulness. Yet every added layer must still meet the same test: does it genuinely restore reference, or does it merely replace missing visual cues with a cleaner display and stronger institutional hope? The horizon is the proper measure. If the system can lose the visible horizon and still preserve a transparent, governable understanding of altitude, drift, obstacle risk and recovery geometry, then poor contrast has been managed. If not, the operation is borrowing certainty it does not possess.

That leaves a harder question than whether the aircraft can technically fly in flat light. Many can. The real question is whether the operation can keep its schedule, judgement and accountability once the world stops offering edges for free. A programme that treats whiteout as weather alone will keep searching forecasts. A programme that treats it as an operational condition will start examining its assumptions about recovery windows, abort criteria, route structure, crew workload and the exact moment at which an image ceases to deserve belief.

Infrastructure scarcity and distance penalties

In January 1925, when diphtheria threatened Nome, Alaska, the serum did not arrive by elegant theory. It arrived by relay: men, dogs, sledges, weather, fatigue, handovers and route discipline across a hostile distance. The drama entered folklore, but the operational lesson is colder and more durable. Remote survival depends less on the brilliance of any one vehicle than on the chain that feeds, shelters, repairs and recovers it. Distance keeps the bill.

That old arithmetic survives beneath modern language about autonomy. An unmanned aircraft can spare a pilot the danger of low-visibility flight. An uncrewed surface vessel can remain on station without bunks, a galley, or watch rotation. A remote sensor can outlast the patience of a human observer. All true. Yet the Arctic imposes its own correction. Machines do not free an operation from its infrastructure; they change the kind of infrastructure it requires. Instead of crew quarters, they ask for heated storage, clean charging, spares, data handling, recovery plans, maintenance shelter, trained hands, and some means of reaching the machine when it ends up somewhere inconvenient. The celebrated absence of a person on board often conceals a growing presence everywhere else.

This is where the idea of infrastructure has to be widened beyond runways and ports. In severe northern operations, infrastructure means any arranged support that preserves repeatability. A hangar, yes, but also a heated container, a generator that starts reliably, a bench with proper lighting, dry storage for batteries, bandwidth sufficient to move logs and imagery, fuel that can be delivered on time, packaging that survives transit, and enough spare parts in the right place that a small defect does not idle an expensive system for want of a seal or connector. Show the matter plainly, and the concept becomes harder to romanticise. A drone may complete a mission flawlessly, then sit grounded for days because the only sheltered work area is occupied, the correct replacement part is far away, or the data needed to diagnose a fault cannot be moved over a thin backhaul. Name the condition and it becomes more exact: logistical dependency. Explain it and the implication is blunt. Autonomy does not remove support. It redistributes it.

The Arctic punishes anyone who treats that redistribution as a clerical detail. Long lines of communication alter the meaning of time. A minor fault in a well-supported base is an inconvenience; the same fault in a remote site becomes an operational event because every remedy takes longer. Parts travel farther. Weather interrupts transport. Skilled maintainers cannot be summoned casually. Cargo priorities compete. Even information moves more slowly when links are patchy or expensive. The consequence is not only delay. It is decision distortion. Teams begin to husband components, defer inspections, stretch intervals, or accept ambiguous diagnoses because the alternative is lost tempo they feel they cannot afford. Risk, in such settings, becomes visible as a maintenance sign-off made with thinner confidence, a sortie cancelled for lack of one trivial item, a recovery deferred because no plausible diversion field can be supported if things go wrong.

The absence of hangarage deserves special attention because it changes the condition of the machine before anyone has chosen to fly it. Cold-weather operations rely on controlled spaces more than temperate planning often admits. Without shelter, snow and moisture accumulate during preparation. Batteries cool while crews work. Tools become harder to use. Connectors and seals are handled in worse conditions. Inspections slow down just as fingers lose dexterity and daylight narrows. A vehicle that lives outdoors between sorties is not simply parked; it is ageing in the open. Show the routine and the point is obvious: a crew uncovers an aircraft in blowing snow, brushes off accumulation, warms what it can, fits a battery, checks surfaces, squints at a display, and rushes slightly because exposure taxes both people and equipment. Name the issue and it is not merely discomfort. It is degraded maintainability. Explain why that matters and the operational penalty comes into view. The sortie begins with less margin than the checklist implies.

Data backhaul creates a parallel constraint, less visible but often just as decisive. Modern unmanned systems promise awareness through sensing, recording and transmission. In remote Arctic operations, those promises meet narrow pipes. Satellite connectivity may be intermittent, expensive, delayed, geographically uneven, or allocated among too many competing needs. A platform can collect useful imagery or telemetry and still fail the mission in practical terms if the data cannot reach the people who need it while decisions remain live. Worse, poor backhaul undermines diagnosis. Fault logs, system health files, video traces and maintenance records become harder to review promptly with distant specialists. The machine may return physically intact yet remain operationally opaque. Efficiency appears here as a very concrete behaviour: whether crews can decide by the next launch window that a fault is understood, tolerated, or disqualifying.

A named modern example makes the point. During the MOSAiC expedition, beginning in September 2019, the German research icebreaker *Polarstern* drifted with the sea ice across the central Arctic under the leadership of the Alfred Wegener Institute. The expedition was scientific rather than military, but it offers a revealing lesson in remote operations. Almost everything depended on organised support: resupply windows, power management, communications discipline, shelter, prioritised maintenance and careful movement of instruments and people in a place where distance and weather narrowed every option. Instruments could be sophisticated; the environment still demanded a logistics mind. The significance lies not in any single device deployed from the ship, but in the plain fact that serious Arctic work remained inseparable from the infrastructure that sustained it, even when that infrastructure had itself to be brought along and defended from cold, darkness and ice movement.

A second example sits closer to autonomous systems proper. NOAA and Saildrone conducted uncrewed surface vehicle operations in and around Alaskan waters during the early 2020s, demonstrating the attraction of long-endurance platforms in places where crewed presence is costly and weather windows are unforgiving. These vehicles showed, compellingly, that autonomy can extend reach. Yet their usefulness still depended on support ashore and afloat: command arrangements, communications, data handling, technical oversight, maintenance planning, and realistic expectations about what happens when a remote platform needs intervention. The persuasive part of the example is not that the vehicles required no infrastructure. It is that they performed valuable work precisely because an infrastructure of planning, support and data custody stood behind their apparent independence. The stack extended beyond the hull.

Distance penalties become particularly severe when a vehicle diverts or lands out. In temperate regions with dense transport links, an off-nominal landing may be irritating, expensive, or embarrassing. In the Arctic, it can redraw the entire mission. A drone forced down at an alternate site may be physically recoverable yet still operationally costly if reaching it requires over-snow travel, another aircraft, a detour by vessel, diplomatic clearance, weather delays, or a larger team than the system was supposed to save in the first place. Show the sequence and the stakes are plain. A platform aborts, takes a safe diversion, and survives. Good. Then comes the question the brochure omitted: who gets to it, with what tools, under what authority, and before what secondary damage sets in? Name the problem and it is recoverability. Explain it and the underlying thesis sharpens. Recovery is not the opposite of failure; it is part of capability.

That same logic reaches deep into spares planning. In harsh environments, parts do not fail as isolated abstractions. They fail in families. Connectors, seals, covers, bearing units, propellers, landing skids, exposed wiring, sensor windows and battery accessories tend to suffer together because the same cold, handling and contamination patterns act on them repeatedly. A serious Arctic operator, therefore, cannot think of inventory as a static catalogue. Inventory is a theory about what kinds of damage a season will produce and how much waiting the mission can tolerate. Trust, in this context, means something measurable. It means a maintainer opens a storage case and finds the likely replacement already there. It means a sortie is delayed for inspection rather than cancelled for lack of a consumable. It means a fault discovered at dusk does not automatically become tomorrow's grounded fleet because the next shipment is uncertain.

Maintenance intervals suffer similar pressure. The nominal interval in a manual reflects assumptions about use, environment and inspection quality. Arctic operations strain all three. Environmental severity may justify shorter intervals. Sparse shelter and poor light may make inspections slower and less complete. Distance from specialist support raises the cost of deferral if an incipient problem matures. Programmes that ignore these pressures often end up in a peculiar form of self-deception: they preserve the letter of the interval while quietly reducing the reliability of what "inspected" or "serviceable" now means in practice. A vehicle signed off in a freezing temporary shelter after a rushed turnaround is not in the same condition as one signed off on a warm bench with full diagnostic access, even if both receive the same entry in a logbook. The missing shed returns here, quietly, as a determinant of truth.

A fair account must admit a limitation. Not every Arctic site is equally austere. Northern Norway, parts of Iceland, some Alaskan locations and settlements in Svalbard can offer comparatively mature support compared with drifting ice camps, remote littorals or improvised expeditionary positions. Better runways, power availability, warehouses, fibre links or regular marine access can soften the harshest penalties of distance. Systems operating from these nodes may look impressively sustainable, and often they are. That matters because it prevents the analysis from dissolving into polar fatalism.

But the limitation does not undo the argument. It refines it. Improved nodes do not abolish logistics; they concentrate and reveal it. A well-supported Arctic base works because people have invested in shelter, power, transport, communications, stock control, tooling and training over time. There is no evidence that autonomy transcends infrastructure. This evidence shows that when infrastructure is in place, autonomous systems can convert it into a reliable service more effectively. The moment operations move away from that node, the old penalties begin to reassert themselves. The range of recovery fields grows. Resupply becomes less frequent. Data movement slows. Decisions once cushioned by spare capacity become harder-edged.

Another distortion enters when organisations confuse platform count with operational availability. In remote conditions, owning more vehicles does not automatically lead to greater persistence. The pace-setter may be charging space, hangar slots, generator output, technician hours, or the rate at which logs can be reviewed and faults cleared. Show the contradiction and the familiar language of force sizing starts to wobble. Four aircraft on paper may yield fewer useful sorties than two, if shelter is too limited to prepare them properly or if a single recurring defect ties up scarce maintainers. Name the issue, and it becomes a capacity mismatch. Explain it, and the consequence is sobering. Numbers at the front end can mask fragility in the support chain behind them.

The same mismatch affects data. Platforms are often purchased for what they can see, sample or record, yet the Arctic repeatedly asks a prior question: what can actually be moved, stored, interpreted and acted upon before the moment passes? A sensor that floods a thin communications link with raw output may look impressive in a trial and clumsy in sustained use. A maintenance regime that depends on remote experts reviewing high-volume diagnostic files may fail if the files arrive too slowly. Here again, logistics sits inside the autonomy stack. Data is not an ethereal by-product. It is cargo, with all the familiar burdens of prioritisation, transit, custody and delay.

The most serious programmes eventually come to a harder but more honest view. They stop treating infrastructure as fixed scenery and start treating it as an active component of system design. That means measuring where maintenance truly happens, where spare parts actually sit, how long recovery realistically takes, what power is available at the edge, which communications are dependable enough for live decisions, and how far a vehicle may range before diversion becomes a support problem rather than a flight problem. The machine's envelope and the support chain's envelope have to be compared, not merely admired separately.

The consequence is slightly austere but liberating. Once logistics is recognised as part of autonomy rather than a burden attached afterwards, many debates become clearer. Endurance claims shrink to operationally honest length. Recovery plans cease to sound ceremonial. Spare holdings stop looking excessive and start looking like control of future time. A heated shelter, a better parts forecast, or a more disciplined data pipeline may matter more to real availability than another increment of nominal range or a more dazzling payload mode. None of that flatters technological glamour. All of it serves actual persistence.

The decisive question, then, is not whether an unmanned system can reach far into the Arctic. Many can. It is whether the support chain can reach just as faithfully behind it: with warmth where warmth is required, with parts before need becomes paralysis, with data fast enough to convert observation into understanding, and with a recovery plan that remains plausible after the neat route has broken. When the distance begins to keep the bill, the true autonomy stack is the whole chain that answers.

Chapter 2

Cold-Weather Reliability Engineering Basics

A vehicle returns from the cold looking almost triumphant. The airframe is intact. The flight log shows no dramatic alarm. The payload brought back useful data. Then, in the shelter, a connector feels slightly loose under a gloved hand, a housing carries a trace of moisture where none should be, and a motor that sounded clean in the air now starts with a faint hesitation. Nothing has failed decisively. That is the problem. The bench says it passed.

Arctic reliability rarely collapses in a single grand gesture. It frays at the edges, then migrates inward. A machine can remain tactically impressive while becoming operationally doubtful, which is far more dangerous than simple weakness because doubt breeds bad judgement. Teams launch one more sortie because the fault seems minor. Maintainers defer one replacement because the spare is scarce. Supervisors accept one ambiguous recovery because the mission still looks recoverable. Under temperate conditions such compromises may survive for a while. In severe cold they accumulate interest quickly.

This is where reliability stops being a background engineering virtue and becomes a governing principle. The Arctic does not care how elegant a control interface looks, how intelligent the mission logic sounds, or how persuasive the demonstration footage appears. It asks a harder question: under repeated exposure, can the system still do tomorrow what it did today, without hidden deterioration, improvised excuses, or widening uncertainty about what has changed? That question moves attention away from spectacle and towards repeatability. It also changes the moral weight of design. A platform that performs brilliantly once and unreliably thereafter is not merely disappointing in this environment. It becomes a liability because the very fact of prior success tempts people to trust a machine whose margins are already thinning.

The central tension lies here. Designers and operators want capability, endurance, sensing reach, low crew exposure and a degree of delegated action that makes remote operations worthwhile. Yet every addition that appears to strengthen capability also introduces more surfaces, more interfaces, more thermal decisions, more software states, more maintenance burdens, and more ways for a small fault to propagate. Reliability engineering is the discipline that keeps ambition from turning into self-deception. Not by dampening it, but by forcing every claim into a form the cold can interrogate. A requirement that cannot survive exposure, handling, cycling and diagnosis was never much of a requirement at all.

That matters because Arctic operations punish fragility in slow, expensive ways. Failure here does not merely mean a lost sortie. It can mean a shortened season, a grounded detachment, a support chain bent out of shape by recurrent defects, or an entire concept of operations quietly scaled back because crews no longer believe the system will behave consistently enough to justify risk. Trust, in this setting, is not an attitude. It is visible conduct. It appears when operators accept the planned route without adding private cautions of their own. It appears when maintainers sign off an aircraft without mentally downgrading the next mission. It appears when supervisors do not need to ask whether the cold has changed some minor mechanism in a way no one has yet measured. Where trust is weak, tempo slows, workarounds multiply, and every routine task grows an unnecessary shadow.

The stakes extend beyond hardware. A reliability problem in the Arctic almost never stays politely confined to the part where it began. Thermal behaviour alters enclosure choices; enclosure choices alter condensation patterns; condensation affects connectors and inspection confidence; connector behaviour affects software interpretation; software responses determine whether a sensor fault stays local or turns into an airworthiness question. The system is judged as one body, even when it fails by inches. That is why cold-weather reliability engineering has to be read as a way of thinking rather than a box-ticking speciality. It links matter, logic and sustainment. It refuses the comforting fiction that a sound airframe can compensate for weak interfaces, or that clever code can rescue a component choice made for convenience instead of endurance.

The path ahead follows that chain from requirement to consequence. It begins where all serious reliability work begins, with the discipline of stating what must remain true if a mission is to be repeatable at all. From there, attention shifts into the architecture of heat and enclosure, then into the small but unforgiving decisions around parts, seals and derating that often decide whether cold-weather claims survive contact with handling and vibration. The argument then moves into software, not as an abstract realm apart, but as the place where faults are either contained or amplified, before arriving at the learning loop created by logs, narratives, inspections and returned components once the field starts speaking back.

A useful way to hold these questions together is to think of a door seal. It is ordinary enough to escape notice when it works. Only when cold stiffens it, moisture creeps past it, or the latch no longer closes cleanly does everyone suddenly understand what depended on that narrow strip of material all along. Much of Arctic reliability works in exactly that register. The decisive things are often not glamorous. They are interfaces, tolerances, margins, transitions and habits of verification. Ignore them, and the programme drifts into ritual confidence. Attend to them, and capability becomes something steadier, less theatrical, and far more useful.

The chapter turns, then, on a deceptively severe test: when a system is asked to perform after cold soak, repeated handling, uneven warming, uncertain contamination and imperfect information, which of its claims still deserve belief, and which were only ever fair-weather promises?

Reliability as a design requirement

On 28 January 1986, the Space Shuttle *Challenger* broke apart shortly after launch from Kennedy Space Center in Florida. The accident belongs to a different world from Arctic drones, but one lesson crosses domains without difficulty. A system can look operational right up to the point where an environmental condition exposes a requirement that was never honestly settled. Cold did not create the programme's pressures, but it revealed them with merciless clarity. The machine flew. The design promise did not.

That is why reliability deserves to be treated not as a pleasing attribute to be measured after the exciting work is done, but as a design requirement from the beginning. Arctic operations make this brutally plain. A vehicle may fly beautifully in a demonstration, track cleanly in moderate weather, and impress decision-makers with its sensor picture, endurance, or ease of control. Then the cold arrives, the handling becomes rougher, the margins thinner, the recovery options fewer, and the platform starts to ask a harsher question: under these conditions, what exactly was the system required to keep doing, and how would anyone know if it could not? That question decides whether autonomy remains a capability or becomes a liability. Five words matter here: what must remain true.

Reliability engineering begins with a refusal to confuse performance with repeatability. Show a machine completing a difficult sortie and most observers will naturally focus on what it achieved. Name the missing discipline and the conversation changes: repeatability under stated conditions. Explain the difference and one sees why Arctic programmes so often disappoint when they scale. Performance asks whether the platform can do the thing once, perhaps even spectacularly. Reliability asks whether it can do the thing again and again without hidden degradation, ambiguous failure growth, or an accumulating burden on maintainers and operators that quietly hollows out the mission. The implication is severe. A tactically impressive system that cannot be counted on to repeat its own success in cold weather may be worse than a less dazzling one, because it invites commitments the support chain and risk posture cannot sustain.

That is not a merely technical objection. It is an operational one. When an organisation believes it owns a dependable capability, it plans differently. It assigns tasks with less reserve. It reduces human backup. It narrows recovery windows. It commits to wider coverage or longer legs. It builds tempo around the assumption that yesterday's success has predictive value. If that assumption is false, the damage spreads quickly. Trust, in this setting, is not a feeling. It appears as a planner shrinking contingency allowances, a maintainer signing off a component without inward doubt, a supervisor authorising launch near the edge of weather or daylight because the system has earned that privilege. Reliability changes behaviour because it changes what people dare to assume.

This is why requirements matter so much. A requirement is not a slogan about ruggedness, intelligence or Arctic readiness. It is a claim precise enough to test and consequential enough to guide design. The serious question is never whether a platform is reliable in the abstract. Reliable for what, under what conditions, for how long, after what exposures, with what maintenance, and with what tolerated degradation? Once framed that way, reliability stops sounding like a soft virtue and begins to look like a contract between environment, design and use. A requirement that says, in effect, the aircraft should function in cold weather means almost nothing unless someone has already decided what counts as function, what range of temperature and handling matters, what mission phases are included, and what failures convert nuisance into mission loss or safety concern.

A named historical example shows how quickly weak requirements can become dangerous illusions. The *Challenger* disaster is often remembered in public shorthand as an O-ring failure. That is true, but insufficient. The deeper failure lay in the mismatch between what the system was expected to endure and how uncertainty around low-temperature seal behaviour was translated, or not translated, into operational limits and design judgement. The requirement space was not honestly aligned with the environmental reality. The mechanical detail mattered enormously. So did the institutional treatment of that detail. For Arctic unmanned systems, the lesson is not that every programme courts catastrophe in the same way. It is that a reliability requirement ignored at the margins does not stay marginal for long. It migrates upward into scheduling, risk acceptance, launch authority and ultimately organisational credibility.

A second example, quieter but no less instructive, can be found in the long history of polar aviation and Antarctic field operations, where aircraft, vehicles and instruments have repeatedly taught the same lesson: equipment that performs adequately in temperate qualification often behaves differently once repeated cold soak, handling in gloves, wind exposure and constrained maintenance become normal. The British Antarctic Survey, among others, has long operated aircraft and field equipment in conditions that reward disciplined preparation over optimistic extrapolation. The checkable fact here is not a single dramatic incident but a pattern of institutional behaviour. Successful polar programmes build reliability into equipment choice, inspection routine, packaging, spares posture and procedural limits. They do not wait for the field to adjudicate vague aspirations. They turn environmental assumptions into explicit operational conditions because the cold will do so anyway.

The governing principle is simple, though not easy. Requirements must translate into testable claims. Show the original design brief for an unmanned platform and it may contain attractive language about endurance, payload, autonomy level, deployability and operating envelope. Name what reliability engineering adds and the language becomes less flattering but more truthful. It asks how often a mission can be repeated before a critical fault becomes likely, what degradation is acceptable before recovery is mandatory, what maintenance actions are assumed between sorties, what component life is expected after thermal cycling, and which sensor or control failures must remain contained rather than propagate. Explain why this matters and the benefit becomes obvious. A testable claim disciplines both enthusiasm and denial. It forces a programme to compare promise with evidence rather than treating field trouble as an unfortunate surprise.

Cold-weather unmanned systems sharpen this discipline because the mission is usually a chain rather than a single event. Launch, transit, sensing, communications, navigation, recovery, post-flight inspection, charging, data handling, and readiness for the next sortie all belong to one reliability picture. A requirement that speaks only to airborne performance is therefore incomplete. If a drone can complete its route but returns with condensation inside a housing, an intermittently loose connector, or a battery whose cold-weather recovery is slow enough to break turnaround assumptions, then the mission succeeded only theatrically. Operationally, the system may be failing already. Reliability engineers think in these longer loops because the field does not grade on isolated moments. It grades on recurrence.

That is where the Arctic strips away flattering language. A platform described as highly capable may still depend on a maintenance tempo, storage condition and spares availability that remote sites cannot sustain. A vehicle marketed as autonomous may still require operator oversight and technical intervention so frequently that its supposed labour savings evaporate in practice. An aircraft praised for endurance may still be a poor Arctic tool if the conditions that preserve reliability force such conservative launch, recovery and inspection discipline that real availability remains low. Efficiency, here, becomes measurable in visible operational terms: sorties launched without extraordinary preparation, recoveries completed without special pleading, defects diagnosed without guesswork, maintenance completed within the support reality of the site, and enough confidence preserved that tomorrow's mission is not treated as a gamble disguised as routine.

The temptation, especially in organisations attracted to novel capability, is to postpone this honesty. They tell themselves that reliability can be improved later, after deployment, after initial procurement, after users gain more familiarity, after software updates smooth the rough edges. Sometimes that is partly true. Fielded systems do improve. Software does mature. Fault data does accumulate. Yet cold-weather operations impose a limit on how much unreliability can be tolerated as a development strategy. The environment is too punitive, the support chain too stretched, and the cost of ambiguous behaviour too high. What could be learned iteratively in a forgiving theatre may become prohibitively expensive in a harsh one. Reliability, then, is not the enemy of innovation. It is the condition that stops innovation from consuming its own credibility.

A fair account requires a limitation. No engineering team can specify every failure, every interaction, every rare combination of cold, wind, contamination, handling error and latent defect. Perfect foresight does not exist, and requirements that pretend otherwise often become bloated, untestable and performative. Some operational learning must occur in the field, because real conditions inevitably exceed the imagination of pre-deployment plans. That limitation matters. It prevents reliability engineering from turning into a fantasy of total control.

But the limitation does not excuse vagueness. It argues for discipline about uncertainty. Good requirements do not promise omniscience; they define what must be true, what can degrade gracefully, what assumptions remain provisional, and what evidence would justify changing the design or the operating concept. That is a humbler and stronger stance. It permits learning without pretending that early ignorance was wisdom. It creates a language for deciding whether a fault should trigger redesign, additional testing, tighter limits or changed maintenance practice. It also protects operators from being made the absorbent layer for design uncertainty. When a requirement is weak, crews compensate. When a requirement is honest, the system carries more of its own burden.

The idea of failure containment belongs here. A reliable design does not merely avoid faults; it prevents local faults from becoming system-level confusion. A sensor can misbehave without corrupting navigation logic. A connector can degrade without producing a stream of misleading status signals. A battery can underperform without luring the control system into brittle behaviour. This is still reliability engineering, not yet a separate software discussion, because it concerns what the system must continue to mean under stress. Show a recoverable defect and one sees the principle at work: the mission aborts cleanly, the logs remain interpretable, the fault path is bounded, and maintainers know where to look. Name the concept and it sharpens: survivable failure. Explain its importance and the Arctic context makes it urgent. Where recovery is hard and visibility poor, systems that fail noisily but legibly are often safer and more useful than those that continue elegantly while becoming untrustworthy.

The most serious question a reader can ask of any cold-weather autonomous platform is therefore not whether it has been tested, because nearly everything has been tested in some sense. The question is what claim the testing was supposed to settle. Was the requirement merely that the vehicle fly after a cold soak, or that it complete repeated sorties with predictable maintenance and no hidden rise in intermittent faults? Was the requirement that sensors operate, or that their outputs remain trustworthy enough to support control decisions in degraded visual conditions? Was the requirement that batteries power the mission, or that energy margins remain sufficiently stable across age, temperature and handling that planners can build real schedules around them? The door seal returns here as a quiet motif. No one notices it in a brochure. Everyone notices it when the room can no longer be kept warm.

This way of thinking has institutional consequences. Procurement language changes. Acceptance trials change. Maintenance reporting changes. Operators learn to describe not only what went wrong, but what prior assumption failed. Engineers stop asking whether the platform met the requirement and start asking whether the requirement described the mission honestly enough to matter. That shift sounds small. It is not. It marks the difference between buying a machine and building a dependable service. In the Arctic, only the second has a future worth planning around.

The unresolved tension sits exactly where ambition meets evidence. Organisations want systems that do more, go farther, demand less human presence and remain affordable to field. Reliability engineering keeps asking a colder question: which of those desires can be turned into testable claims without leaving crews, maintainers and commanders to discover the real limits by trial in the weather? Until that question is answered with unusual honesty, every impressive sortie carries a shadow.

Thermal design and enclosure strategy

In April 2021, NASA flew *Ingenuity* on Mars after months of anxiety about whether a machine so small could survive nights cold enough to threaten its batteries, electronics and structure in Jezero Crater. Mars is not the Arctic, and the engineering cultures are different, yet the underlying problem is familiar. A vehicle does not merely operate in cold; it negotiates with it constantly. The negotiation begins long before movement, sensing or control. It begins with where heat is allowed to remain, where it is encouraged to leave, and what happens when those choices prove inconsistent across a real operating cycle. Warmth must travel carefully.

That is why thermal design is never just a matter of keeping things from freezing. In severe environments, heat behaves like a scarce and unruly resource. Hold too much of it in one place and condensation, battery stress, uneven ageing or maintenance difficulty follow. Lose it too readily and lubricants stiffen, electronics drift, seals harden, and power systems sag at the moment of greatest demand. The machine is not one temperature. It is a map of temperatures, shifting with flight phase, wind exposure, charging state, sunlight, payload use, and the simple fact that some parts create heat while others merely suffer from its absence.

This is the point at which enclosure strategy becomes architectural rather than incidental. Designers sometimes speak as though the enclosure were just a skin wrapped around the real system. In Arctic operations, that view becomes untenable. The enclosure decides what warms together, what cools separately, what moisture gets trapped, what a maintainer can reach with gloves on, what can be swapped in a shelter, and how quickly faults can be diagnosed when time and light are poor. Show the issue plainly, and the idea becomes concrete. A battery compartment mounted close to warm electronics may preserve useful temperature during transit. The same arrangement may also increase the risk of local condensation after shutdown, complicate field replacement, or heat one component at the expense of another. Name the choice, and it sharpens: thermal architecture. Explain it, and the implication becomes difficult to evade. The thermal design is already a maintenance plan in disguise.

Insulation is often the first answer people reach for, and not without reason. Insulation slows heat loss, protects sensitive subsystems from sudden exposure, and can turn limited internal heat generation into a meaningful advantage. A battery pack, sensor cavity or electronics bay wrapped and compartmented intelligently may remain within a functional range far longer than an exposed equivalent. But insulation never merely protects; it also delays exchange. That delay can be helpful during operation and troublesome afterwards. A compartment that stays warm in flight may cool slowly once power is removed, preserving a hidden warm pocket that attracts moisture from colder surrounding air. A surface that appears safe to open may conceal a temperature gradient across connectors, boards or seals that encourages condensation at precisely the wrong interface. Insulation, then, is not a blanket. It is a decision about time.

Heat sinking appears at first to answer the opposite problem. If components generate waste heat, conduct it outward; if one area overheats, spread the load; if hot spots threaten longevity, tie them to structure and external surfaces. In temperate environments, that logic can be straightforward. In severe cold, it turns sly. A well-designed heat path can stabilise a component under continuous operation and still become a liability during startup, shutdown or intermittent use. External structure that serves as a useful radiator in one phase can become a relentless heat drain in another. A metal chassis that looks reassuringly substantial may conduct valuable warmth away from batteries and sensors faster than the thermal budget can sustain. Good heat sinking, therefore, depends on sequence, not just steady state. The question is not only where heat goes, but when.

Active heating seems, by contrast, satisfyingly direct. If a component needs warmth, provide warmth. Battery heaters, lens heaters, anti-condensation elements and warmed bays all have their place, and often a necessary one. Arctic systems that refuse active heating on principle tend to discover that passive cleverness has limits. Yet heating solves one problem by creating several others. It consumes power that cannot also serve propulsion, sensing or communications. It adds controls, wiring, failure modes and software states. It can produce steep local gradients that place stress on adjacent materials. It can lure designers into tighter packaging because warmth appears available, only for maintainers later to discover that access has become miserable and repair times intolerable. A heated enclosure is not simply a warm enclosure. It is a more demanding one.

Compartmentalisation sits at the centre of these trade-offs because it governs which parts of the machine share one another's thermal fate. A vehicle that places batteries, flight computers, payload electronics and navigation sensors in one common volume may gain from pooled warmth and compactness. It may also suffer when one subsystem's thermal needs conflict with another's. Batteries often want protection from deep cold but also dislike excessive heat. Sensors may require tight stability to preserve calibration. Flight control electronics may tolerate wider swings but resent moisture. Once these parts are locked into one thermal neighbourhood, design compromise becomes unavoidable. Separate compartments can solve part of that problem by isolating needs, yet isolation imposes mass, complexity, seals, cable penetrations and maintenance barriers. The door between compartments is never just a door.

A useful historical anchor comes from Antarctic and polar field engineering more broadly, where equipment intended for cold service has repeatedly been redesigned less because it failed to generate heat than because it mishandled where heat and moisture went next. The pattern can be seen across aircraft, field instruments and shelters used by long-standing organisations such as the British Antarctic Survey: warmth alone does not secure reliability if enclosures trap condensation, obscure inspection points, or make routine servicing too awkward under actual field conditions. The lesson is not attached to one famous incident. It is embedded in the persistence of certain design habits. Systems that survive polar use tend to make moisture paths legible, access deliberate, and compartment boundaries meaningful. They are built by people who learned that a dry, reachable connector may matter more than a theoretically ideal internal temperature.

A second example lies in the accumulated lessons of small robotic systems designed for extra-terrestrial cold, including *Ingenuity*. Part of what made the helicopter remarkable was not only that it flew, but that it endured repeated cycles of cold survival and selective heating within a brutally constrained energy budget. The checkable significance here is not that Martian engineering can simply be copied into Arctic drones. It cannot. The significance is that the design treated thermal survival as a governing architecture from the outset, not a peripheral comfort feature. That distinction matters. When thermal logic is central, enclosure geometry, power allocation, operational sequencing and mission ambition are all forced into honest relation with one another. When thermal logic is late, designers patch cold symptoms while leaving deeper conflicts in place.

The hardest of those conflicts often appears as condensation. People readily understand freezing because ice looks like a direct enemy. Condensation is subtler. A vehicle comes in from the cold, is moved into a warmer shelter, and begins to breathe. Moisture forms where temperature differences, enclosed air and surface properties invite it. The danger is not only liquid water itself, though that is enough. It is the sequence that follows: moisture sits unseen, shifts under gravity or capillarity, reaches a connector or board edge, then later freezes again when the system returns outside. A mission can therefore plant the seed of the next mission's fault. Show that sequence and the idea of enclosure becomes more serious. Name the issue and it is not simply weathering. It is internal climate control, however crude. Explain it and the implication becomes operational. A maintainer's time, a drying routine, an access panel arrangement, a vent location, even the order in which components are powered down, may all affect reliability more than the public specification sheet suggests.

This is why field repair must stay in the conversation. Thermal design that looks elegant in a laboratory can become obstructive in a shelter or on a windswept apron. A deeply nested warm bay may protect electronics beautifully while turning a simple battery swap into a fiddly task that requires bare fingers, extra tools or longer exposure of sensitive internals. A highly sealed payload enclosure may keep contamination out yet make inspection slow and uncertain. A compartment wall may improve temperature stability but force technicians to disturb multiple adjacent systems to reach one suspect part. Efficiency appears in very physical form here: minutes spent opening the machine, the number of fasteners removed with gloves on, whether a fault can be isolated without warming the whole vehicle, whether replacement can occur without exposing warm internals to blowing snow. These are thermal questions as much as maintenance questions, because the enclosure created them.

A fair objection deserves room. Some platforms do work well with relatively simple thermal strategies. Larger airframes with generous power margins, stable mission profiles and good ground support can rely more on passive design and procedural discipline than on elaborate heating and compartmentalisation. Fixed installations or vehicles based near mature infrastructure may avoid the worst consequences of awkward enclosure design because time, shelter and specialist help are available. That matters. Not every Arctic system needs an intricate thermal architecture, and over-design can produce its own failures.

But the objection only holds within conditions that support it. Once infrastructure thins, sortie tempo rises, or field access becomes clumsy, modest thermal shortcuts begin charging interest. A platform that can be babied in a well-equipped base may become untenable at a remote site where drying time is limited, power is rationed, and maintenance must be completed before hands and daylight run out. The reconciliation is therefore clear enough. Thermal strategy need not be maximalist. It must be explicit about what support it assumes. Simplicity is a virtue only when it remains honest about the environment in which it expects to survive.

The same honesty should govern the relationship between thermal design and software. Heaters, thermostatic limits, startup interlocks, warm-up timers and sensor compensation routines all mediate how the enclosure behaves in practice. A beautifully compartmented system can still become unreliable if the control logic warms the wrong thing first, reports state unclearly, or allows operators to misread readiness. Conversely, a less-than-ideal enclosure can sometimes be made serviceable by disciplined sequencing and transparent state reporting. Yet software does not abolish architecture; it inherits it. If the physical arrangement traps moisture, demands excessive disassembly, or places incompatible components in one thermal zone, control logic will spend its life apologising for choices already made in metal, polymer and sealant.

The quiet motif in all of this is the warm pocket. Every cold-weather machine develops them: a battery bay holding residual heat, an electronics corner warmer than the rest, a lens housing just above ambient, a cavity that cools slowly after shutdown. Warm pockets are neither good nor bad in themselves. They become good when design and procedure know where they are, what they protect, and what moisture or imbalance they may invite. They become bad when they remain accidental. Much of Arctic reliability turns on whether a system's warm pockets are planned or merely discovered.

Once seen that way, thermal design stops looking like a specialist concern tucked away beneath broader discussions of autonomy. It becomes part of the operational contract. Insulation affects how long a system can wait outside before launch. Heat sinking affects whether performance changes across a sortie. Heating affects power reserves, duty cycles and confidence at startup. Compartmentalisation affects how quickly faults can be reached, dried, isolated and fixed. Even the humble question of where a panel opens becomes entangled with heat retention, moisture ingress and the risk of exposing vulnerable internals during field service. The enclosure decides more than temperature. It decides what kind of machine this is to live with.

That brings the matter back to architecture in its strictest sense: not appearance, but arrangement with consequences. A thermally clever system that cannot be serviced under real conditions may fail operationally even while its components remain within design limits. A maintainable system that sheds heat too freely may survive inspection but falter in flight. A tightly heated enclosure may preserve electronics while undermining endurance through power draw. None of these tensions can be solved by slogans about ruggedness. They have to be measured, compared and chosen.

The serious test, then, is not whether the machine can be made warm enough for one sortie. It is whether its heat, its dryness, and its access can all be managed together across many sorties without turning maintenance into theatre or logistics into punishment. When a designer chooses where the warmth must stay, what else is being chosen at the same moment: a simpler repair, a shorter mission, a drier connector, a heavier airframe, a slower turnaround, a more honest claim? That is the question the enclosure keeps asking, even when the shell is closed and the vehicle looks ready.

Components, connectors, and derating

On 4 June 1996, the Ariane 5 Flight 501 launcher broke apart less than a minute after lift-off from Kourou in French Guiana. The immediate cause was software, not cold weather, yet the deeper engineering lesson belongs squarely to reliability work in hostile environments. A part or subsystem can behave perfectly within one assumed envelope and fail absurdly once reality strays beyond it. The machine does not care that the assumption was convenient. The machine enacts it. Small margins kill quietly.

That is why component choice in Arctic systems is never a procurement detail. It is a declaration of what the design believes about temperature, handling, vibration, moisture, electrical load, maintenance practice and time. A resistor, seal, bearing, connector, cable jacket or voltage regulator does not merely occupy space in a bill of materials. It carries a view of the world. In mild conditions, many such views appear equally plausible. In severe cold, they stop being equal. The part that looked economical, common, or good enough on the bench begins to reveal the hidden cost of its assumptions.

The bench, indeed, is part of the problem. Bench testing flatters marginal parts because it simplifies the life they will later be forced to live. Show the scene plainly and the weakness becomes obvious. A connector is mated once with warm fingers, routed neatly, left undisturbed, then energised in a clean indoor setting. It passes. Name the missing variable and the test becomes less comforting: accumulated stress. Explain what that means in Arctic operations and the real conditions crowd in. The same connector may be mated and unmated repeatedly with gloves, jostled during transport, cooled deeply, warmed unevenly, exposed to vibration, splashed with meltwater, and asked to keep low-resistance contact after the housing around it contracts at a different rate from the contact system within. The implication is severe. A part may satisfy its nominal specification and still be a poor Arctic part.

Reliability engineers describe part selection in terms of margin because margin is what keeps the environment from writing the mission. The idea sounds abstract until it is translated into visible behaviour. A well-chosen component continues operating after cold soak without needing ritual attention. A poorly chosen one demands warming, reseating, retorquing, cleaning, or cautious interpretation of its signals. Trust shows itself here not in language but in habit. When crews begin pressing a connector twice, wiggling a cable before launch, adding unofficial tape or strain relief, or treating one channel's telemetry as slightly suspect, they are performing a field diagnosis of inadequate design margin. The machine is being carried by procedure instead of carrying its own burden.

Connectors deserve special scrutiny because they are where many harsh-environment failures become intermittent rather than decisive. That makes them operationally expensive. A broken structural member is obvious. A failed contact is often not. A connector may pass continuity checks indoors, then drop a signal only under vibration, only at low temperature, or only after a sequence of handling and refreezing. In Arctic systems, that sequence is not exceptional. It is ordinary life. Show the pattern and the mechanism is easy to grasp. A connector body made of one material and a mating shell of another shrink at different rates. Seals stiffen. Contact force changes slightly. A locking feature that felt positive with bare hands is not fully engaged under gloves. Moisture enters, freezes, recedes, leaves behind a minute residue. The circuit remains functional most of the time. Name the condition and it sharpens: intermittent integrity. Explain why it matters and one reaches the real operational consequence. Intermittent faults waste time, corrode confidence, and invite bad decisions because they resist clean diagnosis.

That is why sealing matters, though sealing alone never solves the whole problem. A sealed connector promises protection from moisture, contamination and salt, and in many cases that promise is worth having. Yet a seal is only as honest as its geometry, compression, material behaviour and field handling. In severe cold, elastomers harden, surfaces contract, and the energy needed to mate components cleanly rises. A connector that is technically sealed may still be improperly seated in the field or may carry a damaged interface no one notices until later. The issue is not that sealing fails as a concept. It is that sealing changes the penalty for imperfection. A poorly sealed connector declares its weakness early; a nominally sealed connector handled imperfectly may defer the penalty until after launch, when the fault has become harder to localise and more likely to masquerade as software instability, sensor error or power fluctuation. The latch matters.

A good example of why engineering discipline must overrule complacency lies in aerospace standards practice itself. For decades, space and defence sectors have relied on derating guidance to reduce the likelihood that components fail when real stresses exceed tidy assumptions. NASA's *EEE-INST-002* and related institutional practices, though developed for a different mission class, rest on a simple principle with wide relevance: do not operate parts near their advertised maximum if reliability matters. The relevance to Arctic autonomy is not bureaucratic but physical. Cold weather rarely acts alone. Electrical demand, thermal transients, vibration, startup surges and handling loads arrive together. A regulator, capacitor or connector current path that looks acceptable when used close to its nominal rating in calm lab conditions may become fragile once those combined stresses enter. Derating is not timidity. It is respect for the fact that specifications are rarely written for your exact misery.

This is where component selection and derating meet. Show a power system built around parts chosen for minimal mass and tight cost, each run near its attractive nominal capability, and the design may look efficient. Name the missing discipline and the truth darkens: stress headroom. Explain how Arctic operations consume that headroom and the result is plain. Batteries present lower effective performance in cold. Heaters add draw. Motors can demand more during sluggish startup or gust response. Connectors carry current through colder, stiffer, more vibration-prone interfaces. Sensors and processors may cycle through warm-up states not well represented in simple duty assumptions. The implication is that every component working "within spec" on paper may still be working too close to the edge in practice. Derating creates room for weather, ageing and imperfect handling to exist without immediately becoming failure.

A second named anchor comes from a less dramatic but highly instructive domain: the long-running standards culture around military and aerospace connectors and parts qualification. Families of connectors built to recognised harsh-environment standards did not emerge because engineers enjoyed paperwork. They emerged because repeated service experience showed that civilian convenience parts, however capable in benign use, often underperformed under vibration, contamination and thermal cycling. One does not need to romanticise standards to see the lesson. Qualification matters when the field is allowed to answer back. The Arctic answers back quickly. A connector that survives a short demonstration flight may loosen after repeated handling with gloves and repeated contraction and relaxation of the surrounding structure. The latch that felt secure once may turn ambiguous by the tenth cycle. The part did not betray the mission. The mission simply discovered which life had actually been designed for.

This is also why derating cannot be reduced to a numeric formula applied late in the programme. It is a habit of mind that begins during architecture. A cable chosen for flexibility indoors may become too stiff in cold to preserve strain relief at the connector. A relay, regulator or memory device chosen at the edge of its environmental range may not fail immediately but may age faster under repeated exposure. A bearing or motor selected for nominal torque may become reluctant when the grease thickens and the voltage sags. Derating asks a sharper question than "does it work?" It asks "how much insult can it absorb before the answer changes?" In Arctic systems, that question should govern the whole parts philosophy.

It would be easy at this point to turn derating into a moral slogan for conservatism. That would be a mistake. Over-conservatism has costs, some of them serious. Heavier parts can increase mass and reduce endurance. More heavily rated connectors can be larger, more expensive, and harder to package cleanly. Wide design margins may consume volume, budget and integration effort that other mission needs also claim. The counter-example matters because some successful small systems do rely on commercial or lightly ruggedised parts, particularly when missions are short, support is plentiful, and the environmental envelope is narrow and well policed. Not every Arctic platform requires the same level of severe-environment pedigree in every component.

But that limitation does not rescue marginal design. It simply makes honesty more important. If a programme chooses lighter, cheaper or less heavily qualified components, it must accept the operational contract that choice imposes: tighter weather minima, stricter handling discipline, shorter maintenance intervals, more conservative spares posture, narrower deployment scenarios, or faster replacement of suspect parts. That is the reconciliation. There is nothing sacred about the most rugged component in every location. What matters is that the system does not pretend a narrow margin is a broad one. The moment organisations advertise wide autonomy while quietly depending on delicate parts and careful treatment, the Arctic begins drafting the correction.

One of the more neglected elements in this correction is the effect of glove-handling itself. Engineers often imagine connectors, latches and service points through the dexterity of the bare hand. Arctic operations substitute reduced tactile feedback, bulkier grip, poorer fine control and more hurried exposure. The physical gesture changes. A threaded collar that can be started easily indoors becomes fiddly outside. A push-pull connector that gives a crisp engagement in training may feel ambiguous under thick gloves. A tiny alignment mark may disappear in dim light or blowing snow. None of these are glamorous observations. All are decisive. A connection scheme that depends on delicacy is not merely inconvenient in the cold. It is architecturally unserious.

The same is true of vibration. Many components survive temperature or vibration in isolation and then fail under the sequence the field actually delivers. Transport to the site shakes the system. Cold soak changes material compliance. Launch loads arrive. The mission adds cyclic vibration. Recovery jolts the structure. The platform is carried, packed, unpacked, and reassembled. A contact spring loses a little force. A solder joint, crimp or insert begins to move microscopically. Bench confidence survives because the failure is still embryonic. Then one morning the same channel flickers during a critical phase, and the organisation discovers that what it had treated as a single flight article was in fact a mechanical history accumulating in silence.

The latch returns here as a quiet motif because it captures the larger truth. Good cold-weather engineering often depends on whether small interfaces communicate their state clearly and maintain it through abuse. A latch that is either engaged or not engaged, with little ambiguity, is more than a convenience. It is a form of operational clarity. A connector that tolerates imperfect alignment, repeated handling, cold contraction and moderate contamination without quietly degrading earns real trust because it preserves meaning under stress. That is what the best component choices do. They keep the rest of the system from having to guess.

This extends beyond hardware into doctrine and maintenance. When parts are marginal, organisations compensate culturally. They carry unofficial spares. They teach workarounds verbally rather than writing them down. They begin to identify certain airframes by temperament rather than configuration, as if one vehicle were "fussy" and another "good". Those are not charming field customs. They are evidence that the parts philosophy has slipped below the operational burden. A reliable programme should not need folklore to explain why one connector behaves after warm-up but not after transit, or why a particular payload must be reseated twice to be trusted. Folklore is expensive.

A mature design culture therefore treats component choice, connector strategy and derating as expressions of mission truth. What environment will the system really inhabit? What maintenance quality will it really receive? How many times will it really be opened, carried, mated and checked in poor conditions? What startup loads, thermal excursions and vibration sequences will it really see? Once these questions are faced without vanity, many attractive parts lose their appeal. They were not wrong. They were simply meant for gentler lives.

The unresolved test is mercilessly practical. When the platform has been cold-soaked, shaken, handled with gloves, warmed unevenly, flown, recovered, and prepared again under time pressure, which parts still behave as though the original specification meant something, and which reveal that the design mistook bench survival for field reliability? Until that question can be answered without ritual caution at every latch and connector, the system is still negotiating with its own assumptions.

Software, watchdogs, and fault containment

On 4 June 1996, Ariane 5 Flight 501 lifted off from Kourou and broke apart less than a minute later after software inherited from Ariane 4 failed under conditions it had not been built to endure. The hardware did not suddenly become meaningless. It became unprotected by the logic meant to govern it. A fault in one place travelled too far, too fast, and the system ran out of ways to remain itself. Code can spread damage.

That lesson belongs in the Arctic more than enthusiasts for autonomy sometimes admit. Severe cold, intermittent sensing, vibration, icing residue, drifting calibration and uncertain communications do not merely threaten hardware. They produce the kind of ambiguity in which software either contains trouble or magnifies it. A sensor value arrives late, jitters, freezes, or becomes physically plausible but contextually false. A data bus hesitates. A positioning estimate remains numerically tidy while the world that once anchored it has thinned into low contrast and repeated surfaces. The aircraft still flies. The question is whether the software continues to understand what kind of truth it is being offered.

This is where software design stops being a story about sophistication and becomes one about discipline. Show a modern unmanned system in operation and much of its apparent competence emerges from software: flight control loops, sensor fusion, mission logic, health monitoring, link management, route execution, power handling and operator interface. Name the central problem and it sharpens at once: fault propagation. Explain it and the practical danger becomes plain. A local defect, such as a drifting sensor or intermittent connector, should remain local if the software is designed with suspicion. If it is designed with optimism, that same defect can migrate into attitude error, false mode changes, unstable control response, misleading operator displays or brittle recovery logic. The implication is severe. In Arctic conditions, survivability depends as much on what the software refuses to believe as on what it computes quickly.

A good watchdog embodies that refusal. The term sounds modest, almost domestic, and that modesty is part of its virtue. A watchdog does not make the machine cleverer. It makes neglect harder. In its simplest form, it checks that expected processes are still alive and behaving within bounds; if they are not, it triggers a reset, a fallback or a controlled transition. Show the need for it in plain terms and the design logic becomes obvious. A process handling navigation data hangs because one malformed input or timing anomaly was never anticipated. Without supervision, the vehicle may continue acting on stale assumptions while outwardly appearing stable. With supervision, the lapse is noticed, bounded and responded to. Name the idea and it becomes reliability language rather than folklore: liveness assurance. Explain the implication and one sees why it matters in the cold. The Arctic rewards systems that fail noisily and legibly over those that fail elegantly and opaquely.

Yet watchdogs are only as useful as the transitions they trigger. Resetting a process sounds tidy until one asks what the aircraft is doing during the reset, what state information is preserved, whether the fallback mode is actually safe in current conditions, and how the operator will understand what has just happened. This is where graceful degradation enters. The phrase is used casually in technology circles, but it deserves stricter treatment here. Graceful degradation means a system yields capability in a controlled order so that recoverability survives longer than performance. A drone losing confidence in part of its visual navigation may stop attempting low-altitude precision manoeuvres while retaining a safer transit mode. A payload tracking function may suspend itself while the flight-critical core remains conservative and transparent. A platform may reduce mission ambition, climb, loiter, return or hold rather than continue pretending that a compromised sensor picture is equivalent to a healthy one. Graceful degradation is not softness. It is hierarchy made operational.

That hierarchy matters because Arctic systems are often asked to operate near perceptual limits. A camera in flat light may still produce an image. A range estimate may still update. A GNSS solution may still exist. None of that guarantees that the software should continue treating those streams as equally trustworthy. Show the problem and it becomes familiar to anyone who has watched a machine persist too confidently. The data remains numerically present, so downstream logic keeps consuming it as if presence were equivalent to integrity. Name the flaw and it becomes easier to think about: false continuity. Explain why it is dangerous and the system-level consequence follows quickly. The machine does not need a dramatic failure to become unsafe. It needs only to remain coherent around a lie for long enough to commit to the wrong action.

A striking historical example lies in the loss of Air France Flight 447 over the Atlantic in June 2009. The aircraft was crewed, the environment was not Arctic, and the causal chain was complex. Yet one lesson is highly relevant. When pitot icing led to unreliable airspeed data, automation modes changed, protections altered, and the relationship between human understanding and machine behaviour deteriorated rapidly. The value of the example is not in forcing a simple analogy. It is in showing how mode logic and degraded sensing can interact under pressure. Systems do not merely suffer faults; they change character when their assumptions about input quality no longer hold. Arctic autonomous platforms face their own version of this challenge when visibility collapses, icing degrades sensors, or navigation cues thin out. The software must know not only that information is arriving, but whether that information still deserves authority.

Mode reversion is one of the chief tools for preserving that authority honestly. A system designed for contested or degraded conditions often contains multiple ways of behaving: full mission execution when data is rich, constrained navigation when uncertainty rises, safe hold or route simplification when confidence falls further, and recovery or abort when the cost of continuation becomes too high. The design question is not whether multiple modes exist. Almost all serious systems have them. The question is whether the transitions between them are comprehensible, timely and conservative enough to protect the mission from ambiguity rather than adding to it. An opaque mode change can be as dangerous as no mode change at all, because operators supervising from a distance may not understand whether the aircraft is prudently protecting itself or merely behaving strangely. Trust becomes visible again. It appears as a supervisor who lets the aircraft execute its fallback because the logic has earned confidence, or as one who grabs manual control because the machine has become unreadable.

A second named example comes from NASA's Mars rovers, particularly *Spirit* and *Opportunity*, which depended on fault-protection software to survive communication delays, uncertain terrain interaction and environmental stress on another planet. The mission context was nothing like Arctic field operations, yet the engineering principle is directly useful. Where remote systems cannot rely on immediate human intervention, software must detect anomalies, preserve essential state, and retreat to safe configurations before small troubles become mission-ending ones. The significance of these missions lies not in romantic comparison but in technical honesty. Fault protection was not a flourish added to the interesting part of the rover. It was part of what made the rover an operable system rather than a hopeful machine. Arctic autonomy deserves the same seriousness.

Still, fault containment is not a magic wall. Software can only contain failures it can meaningfully observe or infer, and severe environments produce awkward forms of uncertainty. A sensor can drift within physically plausible bounds. A connector can chatter just enough to create intermittent corruption that resembles transient noise rather than clear breakage. A heating routine can create timing differences that destabilise one subsystem without crossing any dramatic threshold. The limitation matters because it keeps the discussion honest. Not every damaging interaction can be predicted or neatly isolated. Some faults will remain stubbornly ambiguous until post-flight analysis, and some will arrive as combinations too rare to have been seen in testing.

But that limitation does not weaken the case for disciplined software. It strengthens it. Precisely because not everything can be predicted, the system needs habits of suspicion. It needs cross-checks among sensors that fail differently. It needs health monitoring that watches rate of change as well as absolute value. It needs controls on what any one bad input is allowed to influence. It needs clear distinctions between mission-critical and mission-enhancing functions. It needs logs rich enough that a confusing event can later be reconstructed. And it needs an operational philosophy that prefers recoverable loss of capability over unrecoverable persistence. The Arctic does not reward bravado in code.

The hardest failures often occur at the seam between software and the human beings supervising it. A platform may detect degraded confidence correctly, revert to a conservative mode, and still create danger if the operator is not told in language or symbols that match the urgency of the change. The machine knows its own caution; the human sees an unexpected deviation. Show the scene and the stakes become concrete. A vehicle abandons a precision recovery path and climbs or diverts because it no longer trusts altitude cues. The operator, seeing only that the aircraft is no longer doing what was expected, may assume fault, seize control, or command a return to the very behaviour the software has judged unsafe. This is not merely a display problem. It is part of fault containment. If the human supervisor cannot understand degraded mode logic quickly, the system has left one of its most important interfaces undefended.

That is why good software architecture in harsh environments tends to be slightly austere. It resists unnecessary interdependence. It avoids allowing a payload anomaly to threaten flight-critical control. It separates high-consequence functions from those whose temporary loss is tolerable. It records state transitions clearly. It resets selectively rather than theatrically. It makes conservative assumptions about stale data. It prefers bounded retreat to improvised heroics. These choices can look unspectacular beside demonstrations of advanced autonomy. They are more important. A recoverable machine rarely seems glamorous until the weather worsens, the sensor picture grows dirty, and the only reason the sortie does not become an incident is that the software knew how to become smaller without becoming confused.

The Arctic presses especially hard on that last idea. Software in benign conditions may be praised for making full use of every available input, every optimisation, every marginal gain in precision. Software in severe conditions must also know how to stop. It must know when a beautifully fused estimate has become too dependent on correlated weakness, when a camera feed no longer deserves weight in guidance, when a mission can no longer justify the assumptions required to continue, when a lost-link return is safer than a stubborn attempt to complete tasking, and when a subsystem should be isolated rather than endlessly retried. The watchdog is one expression of this discipline, but the larger principle is restraint encoded as behaviour.

Some will object that this sounds like a bias towards conservatism that may sacrifice capability unnecessarily. The objection deserves respect. A system that abandons missions too readily, reverts too often, or distrusts sensors at the slightest anomaly can become operationally timid and economically disappointing. False positives in fault logic waste time and frustrate operators. Overly cautious reversion can make autonomy feel less like an asset than a bureaucrat in code. That is a real danger, and it explains why software teams sometimes drift towards permissiveness. They fear that caution will make the machine seem mediocre.

Yet this is exactly where Arctic survivability demands intellectual honesty. The relevant comparison is not between a bold system and a timid one on a perfect day. It is between a system that preserves legibility under stress and one that preserves style until the moment it does not. In severe environments, capability that depends on pretending degraded inputs are still trustworthy is a false economy. The better question is how much mission value can be surrendered, and in what order, while keeping the aircraft recoverable, the logs interpretable and the operator inside the loop of meaning. A mission shortened cleanly often teaches more than a mission completed by denial.

One quiet motif runs through all of this: the handrail. In a dim stairwell, a handrail does not make the building elegant. It makes a misstep less final. Software fault containment plays the same role. A watchdog, a fallback mode, a bounded reset, a clear reversion path, a conservative integrity check - none of these make the platform dazzling in a brochure. They make it survivable when sensors misbehave, assumptions decay and the world ceases to match the clean conditions under which confidence came easily. The handrail matters most when it is not dramatic.

The unresolved difficulty lies in deciding where software caution should begin. Trigger too early and useful capability is surrendered. Trigger too late and the local defect has already become a system problem. That threshold cannot be settled by elegance alone. It has to be earned from testing, maintenance evidence, operator experience and an unusually candid account of what kinds of uncertainty the platform can actually survive. In the Arctic, where a little sensor dishonesty can travel a long way before anyone can physically intervene, the true measure of intelligence may be the system's willingness to distrust itself soon enough.

Field reliability feedback loops

In January 1986, after the loss of the Space Shuttle *Challenger*, engineers, managers and investigators were forced to confront a familiar and uncomfortable truth: warning signs had existed before the catastrophe, but the system for turning scattered field evidence into decisive action had failed. The lesson has been retold so often that it risks becoming ceremonial. It should not. Machines do not usually collapse from ignorance alone. They collapse when evidence arrives in fragments and no institution is fast, honest or disciplined enough to join them in time. The clue was already there.

That clue is the real beginning of field reliability. A cold-weather unmanned system rarely announces its future trouble in a single dramatic event. More often, it sends small, repetitive signals from different parts of the operation. A maintainer notices unusual drag in a gimbal during post-flight checks. An operator reports that a video feed stuttered only during one particular phase of descent. A returned connector shows minute moisture ingress after a sequence of routine cold recoveries. None of these facts, standing alone, compels redesign. Together, they may describe a failing family already moving through the fleet. The Arctic gives organisations very little time to pretend such fragments are harmless.

That is why a feedback loop matters more than a pile of data. Show the routine plainly and the distinction becomes obvious. One programme records faults because policy requires it, archives the paperwork, and moves on. Another records faults, compares them across sorties and sites, links them to conditions, recovers suspect parts, examines them systematically, and feeds the results back into maintenance, design and operating limits. Name the difference and it becomes institutional rather than clerical: learning velocity. Explain why it matters in severe cold and the implication is blunt. Arctic systems age quickly in the field, not necessarily in calendar time, but in exposure to combined stressors. If learning lags behind exposure, reliability debt accumulates faster than the organisation can see it.

A maintenance log is the most obvious tool in this process, and the most commonly underused. On paper it exists to preserve traceability. In practice its value depends on whether it captures mechanism rather than mood. "Intermittent fault observed" is nearly useless in a harsh environment. "Payload yaw drift after recovery in blowing snow; resistance noted in gimbal movement; no fault reproduced indoors" begins to teach. The difference lies in observable sequence. Reliability engineers need conditions, timing, recurrence, handling context, and the exact point at which the system departed from normal behaviour. A good log does not merely state that something went wrong. It reconstructs what the machine was doing, what the people were doing, and what the environment was doing when the claim of repeatability first cracked. That level of detail is not bureaucratic fussiness. It is how weak signals become comparable.

Operator narratives matter for the same reason, though they are sometimes treated as too subjective to count. That is a mistake. In cold-weather operations, many early signs of degradation are perceptual or procedural before they become diagnostic. An operator may notice that a vehicle "feels slow to settle" in gusty recovery, or that a payload picture "looks slightly sticky" before any discrete fault code appears. Those phrases are not engineering conclusions, nor should they be mistaken for such. They are observations from people close to the machine's changing behaviour. Show the value of them in practice and the pattern becomes clear. A flight log records nominal stability. A maintainer later finds nothing decisive. But three operators across two detachments independently describe a brief hesitation in the same manoeuvre after similarly cold recoveries. Name that properly and it is not anecdote. It is pre-failure texture. Explain its significance and one reaches a practical implication. Systems learn faster when programmes preserve disciplined human observation instead of privileging only what the machine has already learned to confess.

Returned parts analysis completes the loop because it forces speculation to meet material evidence. A suspect connector, bearing, seal, board, motor or housing carries a mechanical history that conversation alone cannot settle. Wear patterns, fretting, corrosion, moisture paths, cracked potting, deformed seals, backing-out contacts, residue around vents, slight witness marks from repeated glove-handling: such details often reveal whether a fault was random, procedural, environmental or architectural. A returned part is a witness. It cannot speak quickly, but it can contradict comforting stories. A programme that swaps components and discards them without analysis is therefore throwing away one of its few chances to distinguish local mishap from fleet tendency.

A useful named example comes from commercial aviation's long-standing use of reliability and maintenance reporting systems, where repeated low-level fault data has often driven changes more effectively than dramatic singular incidents. The broader airline world is not the Arctic, but its discipline offers a checkable institutional lesson. Serious fleets do not rely solely on major failures to decide what matters. They watch recurring defects, deferred maintenance items, repeated removals and no-fault-found returns because those patterns reveal where the system is quietly wasting reliability. For cold-weather unmanned programmes, the relevance is direct. An intermittent connector replacement that keeps recurring, or a gimbal assembly repeatedly returned with ambiguous drag after snow exposure, should not be filed as isolated inconvenience. It should trigger comparison across aircraft, crews, routes and conditions. Pattern beats surprise.

A second example can be drawn from unclassified lessons in polar and remote scientific operations, including work by organisations such as the British Antarctic Survey and the Alfred Wegener Institute. Their equipment portfolios are broader than drones alone, but that is part of the point. Instruments, vehicles and support systems used in severe cold teach reliability lessons only if observations made at the edge can travel back into engineering decisions before another season repeats the same failure. Publicly described polar programmes have long depended on disciplined field notes, defect reporting, and close attention to what returned equipment says about environmental exposure and handling reality. The exact platforms vary. The institutional habit is the real lesson.

Remote operations become safer and more repeatable when the field is not treated as a place where problems merely occur, but as a place where the design continues to reveal itself.

This is where Arctic programmes collide with conventional procurement rhythms. Many acquisition systems are built to favour scheduled reviews, major milestones and formal change processes spaced too widely for harsh-environment learning. That rhythm may function tolerably where equipment sees stable conditions and consequences unfold slowly. In the Arctic it becomes a liability. A winter season can expose enough combined stress to invalidate assumptions made only months earlier. If the mechanism for design adjustment, maintenance revision or operator guidance depends on a slow chain of approvals, the field keeps paying for lessons already half learned. The resulting gap is not abstract. It appears as repeated no-fault-found removals, growing use of unofficial workarounds, expanding caution in crews, and a fleet that remains technically available while becoming operationally mistrusted.

Trust, in this setting, is a measurable social product of feedback quality. Where logs are thin, narratives dismissed and parts analysis delayed, crews start creating their own informal reliability maps. They learn which airframes feel temperamental, which connectors deserve an extra push, which software states they do not quite believe, which weather windows are "probably fine" only on paper. This is often romanticised as field wisdom. Much of the time it is a symptom of institutional lag. The organisation has failed to absorb and formalise what operators and maintainers already know in fragments. A healthy feedback loop converts local caution into fleet knowledge. An unhealthy one leaves reliability trapped in memory, rumour and personal preference.

The speed of that conversion matters enormously. Cold-weather systems do not merely fail under severe conditions; they can drift into unreliability through repeated exposure before anyone notices that the drift has become structural. A component family may show slightly reduced life after thermal cycling. A maintenance task may prove too delicate under gloves and lead to subtle assembly variation. A software threshold may be acceptable in temperate testing and misleading under Arctic sensor behaviour. None of these changes necessarily announces itself as an urgent crisis. They accumulate. Fast feedback turns accumulation into intervention. Slow feedback turns it into culture.

That is why the best field reliability loops are not only technical but narrative. They preserve numbers and stories together. A log file may show when a signal dropped. An operator can describe what the aircraft was doing, what the display looked like, what the wind and visibility felt like, and whether the anomaly resembled previous sorties. A returned part can confirm whether the likely cause lies in ingress, wear, handling, thermal cycling or something more elusive. Each source is incomplete alone. Combined, they become a learning system strong enough to challenge convenience. Show a programme that values only telemetry and it will miss how human supervision actually experienced the anomaly. Show one that values only anecdote and it will drown in impression. Reliability grows where the two are compared, not where one flatters the prejudices of the team.

A fair limitation deserves acknowledgement. Faster feedback is not automatically better feedback. Programmes can overreact to sparse evidence, redesign around noise, tighten limits unnecessarily, or flood teams with defect reports too poorly structured to support meaningful learning. The danger is real. A cold-weather detachment under pressure may attribute every irregularity to the environment and thereby generate a false sense of pattern. Returned parts can mislead if handling after removal alters their condition. Operator recollection can sharpen around dramatic missions and blur around routine ones. Speed without discipline can turn reliability work into superstition with paperwork.

But that limitation does not justify slowness. It argues for structured speed. Programmes need ways to compare like with like, to separate symptom from mechanism, to preserve chain of custody for suspect parts, and to revise maintenance or operating assumptions without waiting for a crisis grand enough to satisfy bureaucratic drama. In practical terms, that means defect categories that reflect real failure families, maintenance logs that record sequence and context, operator debriefs that invite concrete observation, and engineering review mechanisms able to act on patterns before another season makes them expensive. The goal is not to react to everything. It is to make sure nothing keeps recurring merely because the organisation moves more slowly than the cold.

The most revealing sign of a mature feedback loop is that it changes behaviour at several levels at once. A returned connector showing moisture ingress may not only alter the connector design; it may change packing procedures, inspection routines, training emphasis and spare holdings. Repeated operator reports of degraded gimbal smoothness after certain recoveries may not only prompt component review; they may alter wash-down practice, recovery geometry or the sequencing of post-flight checks. This layered response matters because Arctic failures rarely belong to one cause. They emerge from the meeting point of design, environment and use. A learning system that insists on one culprit will learn slowly.

The quiet motif running through all of this is the notebook. Not the romantic notebook of the explorer, but the practical one: the place where one unusual vibration, one awkward recovery, one suspicious trace of moisture, one oddly recurrent reseat of a connector stops being forgotten and enters a chain of comparison. The notebook matters because memory flatters success and erases sequence. Reliability work depends on sequence. When did the drag begin? After what exposure? On which aircraft? Under whose handling? With what follow-on behaviour? The notebook, literal or digital, is where the machine starts to become legible across time rather than merely inspected in the present.

This has consequences for authority as well as engineering. Programmes that learn well allow evidence from the edge to move upward without being polished into harmlessness. Maintainers must be able to say that a component is serviceable by formal criteria and still suspect by lived pattern. Operators must be able to describe degraded behaviour without being accused of imprecision merely because the logs do not yet prove it. Engineers must be able to revise assumptions without treating every field revelation as an embarrassment. The cold exposes systems. It also exposes institutions that prefer tidy narratives to inconvenient mechanism.

The hardest part is deciding when the evidence is sufficient to act. Act too soon and design churn, maintenance burden and operational caution may grow out of proportion to the signal. Act too late and the same weak signals harden into recurring defects, eroded trust and lost capability. That judgement cannot be mechanised entirely. It depends on the quality of the loop itself: how well logs are written, how honestly narratives are heard, how carefully parts are examined, and how willing the organisation is to admit that a tactically successful system may already be operationally drifting. The decisive question is therefore not whether the programme collects data. Almost all do. It is whether the next cold season will begin with the same notebook full of clues, or with evidence already converted into changed design, changed maintenance and a machine less inclined to surprise the people who depend on it.

Chapter 3

Power, Batteries, and Thermal Budgets

A vehicle sits ready on the snow, its airframe clean, its route loaded, its sensors awake. The display says the pack is charged. The crew knows better than to relax. A battery in severe cold is not a container full of obedient energy waiting to be spent. It is a chemical negotiation already under strain, and the first hard demand - take-off, climb, a burst of heating, a sudden correction in wind - may expose how little of that apparent reserve can be trusted. The charge is not the story.

That is the central unease of power in the Arctic. Energy appears measurable, even tidy, and then behaves like something far more conditional. A percentage on a screen suggests certainty. The machine itself may suggest the same: systems nominal, voltage acceptable, mission ready. Yet cold changes what stored energy can actually deliver, when it can deliver it, and at what cost to the rest of the sortie. A battery pack does not fail only when it is empty. It can become operationally poor while still appearing technically alive, especially when age, temperature, heating demand and peak load begin to interact. The number lies politely.

This makes power a more exacting subject than endurance claims usually admit. In easier environments, teams can treat energy as a planning quantity: load the route, estimate consumption, add reserve, launch. In severe cold that arithmetic grows unstable because the battery is not merely carrying the mission. It is also struggling to preserve its own useful condition. Warmth must sometimes be bought from the very resource whose duration the mission depends upon. A platform can therefore become more reliable by becoming less ambitious, which feels perverse until one accepts the chemistry and stops treating the pack as a passive fuel tank.

The real tension lies there. Operators want duration, payload performance, heating margin, safety reserve and clear rules for autonomy, all at once. Designers want manageable weight, acceptable cost, practical charging and a system that does not frighten crews with unpredictable behaviour near the edge of depletion. But every concession to one demand presses on another. Extra heating may protect the pack while shortening the sortie. Conservative return logic may preserve recoverability while reducing useful time on task. Faster turnaround may satisfy operational appetite while asking too much of charging conditions or battery handling in the cold. Power is never only electrical. It is architectural and procedural from the start.

That is why the stakes extend beyond whether a vehicle can stay airborne for a respectable period. A weak understanding of cold-weather power becomes visible in much smaller, more consequential ways. It appears when crews quietly shorten routes because they no longer trust the displayed reserve during climb-out. It appears when maintenance teams spend precious shelter time warming packs before installation. It appears when a platform aborts not because it lacks any energy at all, but because its remaining energy can no longer be delivered with the steadiness required for safe recovery. Efficiency, in this setting, is not a brochure claim about maximum endurance. It is the number of missions completed without surprise behaviour, emergency curtailment, or a growing backlog of packs whose health no one is prepared to certify with real confidence.

The issue lasts beyond the sortie because power in remote conditions is always a chain. Batteries must be charged somewhere, by something, under conditions that may vary more than design documents politely assume. Generators, shore power, heating shelters, connectors, charge routines, transport cases, warming delays and waiting queues all start to matter. A pack that is technically usable may still be unavailable because the site cannot restore it quickly, safely or predictably enough. A platform advertised as persistent may therefore reveal a strange dependence on the dullest objects in the detachment: cables, sockets, fuel cans, extension runs, drying benches, charging logs. The battery leaves the aircraft and enters logistics.

The path through this chapter follows that pressure as it moves from chemistry into operations. It begins with what cold does to stored energy when the mission starts asking for real work, then turns towards the thermal bargains by which warmth is preserved or purchased inside the system. From there the argument passes into charging, where remote infrastructure starts dictating sortie rhythm more openly, before reaching the software and doctrine that decide how a machine should behave when its energy state is no longer generous enough to support optimism.

A useful way to hold these questions together is to think of a thermos flask. Fill it well and it buys time, but not indefinitely. Open it carelessly, leave it exposed, or depend on it to do more than it was built for, and the illusion of abundance fades quickly. Battery systems in the Arctic behave with a similar sternness. Their value lies not only in how much energy they contain, but in how honestly the whole operation understands the conditions under which that energy remains available, recoverable and worth trusting. A warm pack, like a warm drink, is often more useful than a larger cold one that promised more and delivered less.

The harder question, then, is not how much energy the platform carries at launch. It is what share of that energy can still be counted on after cold soak, heating demand, load spikes, charging compromises and the small indignities of field handling have already taken their share - and whether the machine, its operators and its doctrine all agree on that answer before the weather forces the issue.

Battery chemistry and cold performance

In April 2021, NASA flew *Ingenuity* in Jezero Crater after months of concern about whether a battery-powered aircraft could survive repeated cold Martian nights and still deliver enough power to fly. Mars is not the Arctic, but the engineering lesson is close enough to be useful. A battery does not simply contain energy. It contains energy under conditions. Change those conditions sharply enough, and the same pack becomes a different machine. The gauge still glows.

That quiet mismatch between apparent charge and usable power is one of the most dangerous misunderstandings in cold-weather autonomy. Operators tend to look at a battery as though it were a tank: fill it, measure it, spend it. Chemistry does not oblige that metaphor. Cold slows ionic movement, increases internal resistance, changes voltage behaviour under load, and narrows the difference between a healthy-looking pack and a disappointing one. A battery in severe cold may still report a respectable state of charge while losing the ability to deliver power cleanly when the mission asks for it most. Stored energy is not the same thing as accessible energy.

That difference matters most at moments of demand rather than duration. A platform loitering gently may seem comfortable enough. Then it launches, climbs, fights gusts, powers heaters, or carries a de-icing burden, and the battery is suddenly asked not merely whether it has energy left but whether it can release that energy fast enough without a damaging voltage sag. The distinction has a name that should be treated with more respect in operational planning: power delivery. Explain it in plain terms and the point becomes simple. Capacity tells part of the story. Internal resistance often tells the rest. When resistance rises in the cold, the pack wastes more effort inside itself just as the aircraft begins asking for more effort outside. That is how a battery disappoints dramatically before it is empty.

The chemistry behind this is familiar to battery engineers and often underappreciated by everyone else. Most modern unmanned systems rely heavily on lithium-based chemistries because they offer strong energy density and useful cycle performance. But lithium cells are not indifferent to temperature. Low temperatures slow electrochemical reactions and reduce ion mobility in the electrolyte and electrodes. The cell becomes more reluctant. That reluctance appears as lower effective capacity and poorer voltage stability under load. Show the effect operationally and it is easy to recognise. A pack that delivered a comfortable climb profile in temperate testing now sags sooner during take-off. A vehicle that usually has generous reserve after transit begins returning with the margin uncomfortably compressed. A heater or anti-icing load that seemed modest in planning now competes more directly with propulsion because the battery is already energetically stiff. Chemistry becomes flight behaviour.

Internal resistance is where this stiffness becomes operationally unforgiving. The term sounds clinical, yet its effects are tangible. Higher internal resistance means more voltage drop when current demand rises. It also means more heat generated within the cell during discharge, though not necessarily in a way that helps at the moment it is most needed. A battery under high load in cold weather is therefore pulled in two directions at once: it is less willing to deliver power, but the act of delivering power can also warm it somewhat as current flows. That can create a dangerous illusion. A sortie may appear to improve after the first demanding phase because the pack has warmed slightly, encouraging the belief that the mission is now comfortably inside the envelope. Sometimes that is true. Sometimes it merely postpones the real limit until the next high-load event, often when margin is smaller and recovery more urgent.

This is why take-off and climb deserve special attention. These are not just the opening motions of a mission. They are diagnostic moments in which the battery is asked to reveal what the cold has done to it. A vertical or short-field launch, a heavy fixed-wing departure, or a climb through turbulent low-level air all impose high current draw. If the pack is cold-soaked, aged, or unevenly warmed, that demand can expose weaknesses instantly.

A system that clears the ground and behaves normally afterwards may still have consumed a disproportionate share of its useful power in those first minutes. Mission planning that treats the early phase as energetically ordinary is therefore planning against the chemistry rather than with it.

A named example from aerospace history helps frame this more sharply. During extra-terrestrial robotics missions, including work on Mars by NASA, energy budgeting has long been governed by environmental temperature as much as by nominal battery capacity. The details differ from Arctic air operations, but the principle is exact. A battery-supported system in severe cold does not ask only "How much charge remains?" It asks "At this temperature, after this exposure, with this load profile, how much of that charge can be used safely and predictably?" The value of the example lies not in borrowing Martian romance for polar engineering, but in showing that serious programmes treat chemistry as a mission constraint from the outset, not as a footnote appended to endurance claims.

A second, more terrestrial example appears repeatedly in cold-weather drone operations conducted by research groups, utilities and public agencies using lithium-polymer battery packs. The details differ by platform, but the pattern is consistent enough to be familiar in technical notes and field reports: reduced endurance in cold is only part of the problem. More serious are abrupt drops in available power during high-demand phases, especially when packs are launched cold or cycled repeatedly without controlled warming. The pack that survives a bench discharge may still misbehave in the field because the mission asks for sharp current at moments the chemistry dislikes. That distinction should change how one reads battery specifications. A discharge curve measured under stable indoor conditions is not a promise made to an Arctic climb-out.

Age makes all of this harder. Batteries do not merely lose capacity as they cycle and calendar time passes; they often become less forgiving under load. Internal resistance tends to rise with use, storage history and degradation within the cells. A pack that once shrugged off a cold launch may later show visible sag in the same conditions even if its state-of-charge display appears respectable. This is where field judgement often gets into trouble. Operators naturally think in terms of percentages and remaining flight time. Age changes the more important question: how forcefully can this battery still act? A tired battery may resemble a tired person. The energy seems present until the moment a staircase appears.

That is why energy should be treated as a variable constraint rather than a fixed resource. A mission does not leave the ground with "forty minutes" in any meaningful universal sense. It leaves with a chemical system whose usable performance depends on present temperature, recent handling, internal condition, expected load spikes, and how much uncertainty the operation can tolerate. Show the practical consequence and the planning change becomes obvious. A route that looks feasible for a warm new pack may be marginal for an older pack after exposure on a windy launch site. A reserve that appeared generous in calm flight may become thin once de-icing, climb repetition, or gust response enters the picture. Energy planning, then, is not accounting alone. It is an argument about the battery's present character.

De-icing and heating loads sharpen this argument because they consume energy while also trying to preserve the conditions under which energy remains usable. This is one of the Arctic's most punishing trade-offs. A platform may need battery warming, sensor heating, or anti-ice functions precisely because cold has already degraded the performance of the systems that support safe flight. Yet every watt spent on those functions is unavailable for propulsion or payload. Worse, the load may arrive during phases when current draw is already high. An aircraft lifting into icing conditions, demanding strong climb performance and powering heated surfaces at once is asking the battery to solve several problems simultaneously. If the pack is warm and healthy, the trade may be acceptable. If not, the problem compounds. Energy spent to preserve reliability can itself shorten the mission enough to challenge its value.

Still, a fair account needs a limitation. Not every battery chemistry behaves identically, and not every cold-weather platform suffers equally. Certain lithium chemistries tolerate cold and high-rate discharge better than others. Some systems are designed around careful thermal management, modest discharge rates, or mission profiles that avoid the most punishing load spikes. Fuel-powered aircraft with electrical subsystems can shift part of the burden away from batteries. Larger platforms may carry enough thermal mass and energy reserve to soften the sharpest cold-weather penalties. It would be lazy to suggest that all battery-powered Arctic operations are equally fragile.

But the counter-example does not overturn the main point. It clarifies it. Where cold-weather battery performance proves acceptable, it is usually because the system was designed and operated with explicit respect for chemistry, not because chemistry ceased to matter. Better cell selection, intelligent pack design, pre-heating, load management, conservative reserve rules, and realistic ageing assumptions can materially improve outcomes. The reconciliation is therefore straightforward. Cold does not make battery use impossible. It makes wishful battery use expensive.

That expense becomes visible in mission doctrine. A programme that understands cold performance will ask different questions before launch. Not merely whether the battery is charged, but whether it is warm enough, rested enough, healthy enough, and matched to today's load profile. Not merely whether the planned route fits the nominal endurance figure, but whether it fits the likely power draw during take-off, climb, wind correction and recovery. Not merely whether reserve exists on paper, but whether reserve remains meaningful after the battery has already surrendered its easiest performance to the environment. Reliability appears here as behaviour: a shorter route accepted without argument, an older pack held back from demanding profiles, a recovery started earlier because the current draw trend says more than the displayed percentage.

It is worth noticing how often battery failure in the cold masquerades as something else. A sluggish climb can be blamed on wind. A control anomaly during ascent can be mistaken for tuning. A conservative autonomous return can feel overcautious rather than chemically necessary. A forced early landing may be narrated as bad luck. Yet many such events are the battery speaking through the rest of the system. The chemistry does not use words. It uses voltage, sag, hesitation, slower response, unplanned reserve loss, and the narrowing of once-comfortable margins. Programmes that learn quickly become skilled at hearing that language early.

The quiet motif here is the gauge. A gauge reassures because it turns an invisible process into a simple visible claim. But cold-weather battery operations teach the limits of that comfort. A number on a screen is not false exactly; it is incomplete. It does not fully show the burden of internal resistance, the penalty of age, the cost of warming, or the violence of a short high-load demand. The gauge is useful. It is not sovereign. Serious Arctic energy planning begins when people stop asking only how much charge remains and start asking what the pack can still *do*.

That shift matters because autonomous systems are especially vulnerable to the false neatness of battery displays. Humans supervising at a distance tend to trust tidy indicators more than messy physical reality. If the aircraft reports healthy charge, the temptation is to believe the mission remains energetically sound. Yet the harder truth is that power failure in the cold often begins as a mismatch between displayed sufficiency and actual deliverability. Autonomy features built on battery state must therefore be judged not by elegance alone but by whether they honour chemistry honestly under load. A dynamic return threshold that ignores cold-induced power loss is not truly adaptive. It is numerically confident in the wrong language.

This is why the most serious operational question is not how long the battery lasts in favourable conditions. It is whether the pack can still deliver what take-off, climb, heating and recovery will demand after cold soak, repeated cycling and ordinary ageing have already taken their share. Until that question is faced directly, every Arctic sortie launched on a pleasing percentage carries a small fiction inside it: that stored energy and usable power are still the same thing.

Heating, insulation, and energy trade-offs

In April 2021, when NASA flew *Ingenuity* in Jezero Crater, one of the least glamorous and most decisive questions was not whether the little helicopter could lift from the Martian surface, but whether it could keep its batteries warm enough to remain useful after the cold had done its work. Mars is not the Arctic, and the mission architecture was far more controlled than any field detachment in the High North. Still, the engineering problem is close enough to be instructive. A battery can spend energy merely to stay ready to spend energy. Heat costs flight time.

That sentence carries more of Arctic power planning than many endurance charts do. People tend to imagine heating as a support function added around the edges of the real mission: keep the pack warm, keep the optics clear, keep the electronics within bounds, then get on with the useful work. Severe cold refuses that hierarchy. Heating becomes part of the useful work because, without it, the battery may lose the ability to deliver power cleanly under load. Yet the power spent on warming is unavailable for propulsion, sensing or communications. The more faithfully a system protects itself from the cold, the more openly it competes with its own mission.

This is the paradox that gives cold-weather power planning its distinct severity. A machine can become more reliable by becoming less enduring. Show the practical choice and the point sharpens. A crew launches with a well-warmed battery, conservative heating set-points and a shortened route, then returns with a comfortable reserve and a platform that behaved predictably throughout. Another crew, using the same aircraft, stretches for the longer sortie, reduces warming to preserve nominal endurance, and later encounters voltage sag during a demanding phase of flight. Name the issue and it becomes more exact than simple caution: thermal budgeting. Explain why it matters and the implication is hard to avoid. In Arctic operations, extra minutes in the air may be purchased by consuming the very thermal margin that makes those minutes safe.

Battery warming lies at the centre of this bargain because cold does not merely reduce available energy; it changes how confidently that energy can be delivered. A warmed pack often behaves not like a luxury but like a different class of battery altogether. Internal resistance is lower, voltage response is steadier, and high-load phases become less punitive. That improvement is operationally meaningful. Take-off becomes less of a chemical gamble. Climb performance becomes more predictable. Recovery reserve becomes less likely to collapse at the end of the sortie, when an aircraft may need a final burst of authority against wind or an unplanned go-around. Warmth, then, is not comfort. It is usable power translated into dependable behaviour.

But warmth has to come from somewhere, and that "somewhere" is where the trade-offs begin to bite. In a battery-powered system, active heating usually draws from the same energy store that must also support propulsion and payload. A heater may protect the pack's performance while directly shortening endurance. It may also allow a smaller reserve to remain meaningful by preserving power delivery later into the mission. Both statements can be true at once. That is what makes this subject so resistant to lazy thinking. An aircraft can lose mission time by running heaters and gain mission assurance by doing so. The serious question is not whether heating costs energy. It plainly does. The question is whether the energy spent warming the pack buys back more operational value than it consumes.

Insulation complicates the picture in a subtler way. In ordinary discussion insulation sounds like a free kindness, a passive means of keeping heat where it belongs without further power draw. Sometimes it approaches that ideal. A well-insulated battery bay can preserve the benefits of pre-heating, slow heat loss during staging, and reduce the penalty of exposure on the launch line. Yet insulation is not neutral. It adds mass. It changes temperature gradients. It can trap unwanted heat in some phases and preserve unwanted cold in others if warming is uneven. It can also make maintenance and inspection more awkward, particularly in confined systems already crowded by wiring, mounts and seals. Show the system after a sortie and the implications become practical: a battery compartment that stays thermally stable may also cool slowly, retain moisture longer, or take more time to access for checks and swaps. Insulation buys time, not immunity.

That distinction matters because Arctic missions unfold in phases rather than in a single continuous state. A pack may be warmed in shelter, carried into cold air, held on the launch line, discharged hard in take-off and climb, stabilised during cruise, then exposed again during recovery, shutdown and post-flight handling. Heating and insulation act differently in each of these phases. A design that looks efficient on a steady-state thermal model may behave awkwardly in operations because the important question is not only how much warmth can be held, but when it is available, when it is spent, and how fast it is lost under wind and handling. The aircraft lives in episodes. So does its thermal budget.

A useful named example comes from cold-weather spacecraft and planetary robotics, where thermal survival and mission duration have long been inseparable. Mars rovers and helicopters, including those operated by NASA, routinely treat battery heating as a mission-defining constraint rather than a technical footnote. The public fascination tends to settle on movement, imagery and exploration. Engineers spend much of their attention on where the energy goes before movement even begins. The relevance to the Arctic is not that a drone in Svalbard behaves like a rover in Jezero Crater. It is that serious programmes do not separate thermal management from mission logic. They understand that a watt spent preserving battery competence may be as decisive as a watt spent moving the vehicle.

A second example, closer to field reality in the High North, appears across publicly described drone operations by research teams, utilities and inspection operators working in winter conditions. The recurring lesson is not merely that cold shortens endurance. It is that warmed batteries often make shorter sorties more operationally useful than longer nominal sorties flown with colder packs. A team choosing to pre-warm thoroughly, accept a reduced mission radius, and return with orderly reserve is not necessarily surrendering efficiency. It may be buying consistency across multiple launches, reducing the chance of brownout during climb or recovery, and preserving battery health over repeated cycles. The choice looks modest in isolation. Across a week of operations, it can define whether the programme produces dependable service or a trail of anxious improvisation.

Brownout deserves a more exact place in this discussion because it is where the energy trade-off becomes visible to people who are not electrochemists. A brownout is not the elegant exhaustion of a battery used exactly as planned. It is an untidy collapse of usable voltage under load, often at a moment when the system still appears to possess some energy in reserve. That gap between apparent reserve and real deliverability is what makes heating decisions so consequential. A warmer pack may show a lower nominal endurance figure because energy has already been spent preserving temperature, yet it may be far less likely to brown out when the aircraft asks for peak current. A colder pack may promise more minutes and deliver less authority when those minutes become demanding. Reliability, here, is measured in the plainest possible way: whether the aircraft can still do the hard thing near the end.

This becomes especially acute when other loads join the contest. Payloads consume power, sometimes steadily, sometimes in bursts. Communications links may draw more during certain phases. Anti-icing or de-icing systems, if present, may turn from optional to essential with little warning. Sensors can need their own heaters or stabilisation routines. The thermal budget is therefore not a separate column in an engineer's spreadsheet. It is the argument over who gets to use limited power at the exact moment each subsystem claims necessity. A high-resolution payload may look like the mission's purpose, but if feeding it compromises battery temperature or reserve to the point that safe recovery becomes uncertain, the payload has ceased to be a mission asset and become a bargaining error.

It is tempting to answer this by seeking a single superior solution: more insulation, better heaters, larger packs, lighter payloads, stricter routes. Real systems rarely permit that luxury. Larger packs add weight. More heating reduces endurance. Better insulation can complicate service. Lighter payloads may reduce mission value. Shorter routes preserve reserve but shrink operational reach. The serious work lies in choosing which penalties are acceptable and which are not. That is why thermal strategy is inseparable from doctrine. A programme that claims long endurance in the Arctic without specifying the thermal assumptions behind that claim is not describing a capability. It is merely naming a preference.

A fair counterpoint deserves space. Not every system needs aggressive battery warming, and not every cold-weather sortie turns heating into the dominant constraint. Some chemistries tolerate low temperatures better than others. Some platforms generate enough internal heat during operation to stabilise the pack after launch. Some missions are short enough, or staged from sheltered environments, that passive retention and disciplined handling suffice. Hybrid or fuel-powered systems can shift part of the burden away from batteries altogether. It would be lazy to imply that all Arctic operations hinge equally on electric heating.

But the limitation only sharpens the broader lesson. Where active heating proves unnecessary or modest, that is usually because the mission profile, platform design and support conditions have already been aligned in ways that reduce thermal strain. Shelter matters. Turnaround timing matters. Exposure before launch matters. Battery size relative to power demand matters. The absence of elaborate heating is often a sign that thermal decisions were made earlier and more honestly, not that thermal decisions no longer count. Simplicity, in this domain, is often earned.

What matters most is that organisations learn to treat warmth as part of the mission budget rather than as invisible housekeeping. That changes how sorties are judged. A short mission completed with warm batteries, clean voltage behaviour and orderly reserve may be a stronger expression of operational competence than a longer mission that ends in a hurried return and an unhealthy pack. Efficiency begins to look different under that light. It becomes the repeatable generation of safe sorties, not the one-off extraction of the longest possible flight from a chemically reluctant system. Trust appears in visible choices: accepting a shorter route without embarrassment, delaying launch to warm packs properly, declining to stack payload and heating demands too aggressively, or retiring ageing batteries from the most power-intensive tasks before they force the issue in flight.

Age, indeed, makes the thermal bargain less forgiving. As batteries degrade, they tend to lose not only capacity but also composure under load. Internal resistance rises, the usefulness of heating may become more pronounced, and the penalty for skipping thermal preparation becomes steeper. A battery warming plan that seemed optional with a fresh fleet may become mandatory later, not because doctrine turned cautious, but because the chemistry has changed. This is one reason thermal budgets should never be treated as fixed. They drift with season, pack age, maintenance quality and mission pattern. A system that does not revise its assumptions accordingly may keep flying while quietly becoming less honest about the energy it can count on.

The quiet motif running through all of this is the blanket. A blanket does not create warmth from nothing; it helps decide where warmth remains, how long it lasts, and whether it reaches the place that needs it most. But a blanket can also be too heavy, awkward to move, or poor at solving the deeper cause of cold. Battery heating and insulation behave similarly. They are never just protective accessories. They shape the life of the entire system: what can be powered, how long a sortie remains credible, how quickly a battery can be turned around, how easily the pack can be inspected or replaced, and whether the aircraft reaches the difficult parts of the mission with enough honest reserve to remain governable.

This has direct implications for autonomy as well. An energy-aware aircraft that sheds payload functions, shortens mission ambition, or returns early because thermal demands have altered battery reality is only useful if operators understand the thermal bargain built into those decisions. Otherwise a prudent machine will look capricious. A system that uses heating aggressively may appear to be "wasting" endurance unless crews know that the endurance sacrificed was never safely available in the first place. Transparency matters because thermal trade-offs are counter-intuitive. The right choice can look like self-imposed weakness until one sees the alternative clearly: a longer sortie flown on colder batteries, thinner voltage margin and a greater chance that the final demanding phase becomes a forced landing rather than a controlled return.

The hardest question, then, is not whether heaters shorten missions. Of course they do. It is whether the mission time they take away was ever genuinely safe to promise. In severe cold, a short sortie on warm batteries may be the more ambitious act, because it commits to repeatability rather than theatre. The real test comes when planners must choose between the flattering number and the honest one: the longer route made possible by optimism, or the shorter route that leaves the aircraft enough warmth, authority and reserve to survive the moment the weather asks for one hard effort more.

Charging in remote conditions

In September 2019, when the German research icebreaker *Polarstern* entered the central Arctic for the MOSAiC expedition, almost every scientific ambition onboard depended on something less romantic than discovery: power that could be generated, managed, distributed and trusted in a place where replacement, repair and resupply were slow. The expedition was not a drone detachment, and that is partly why it is so instructive. Remote operations do not begin with the machine in the air. They begin with the socket.

Charging in the Arctic is often treated as the quiet, administrative interval between flights, the time when batteries move from used to ready and crews prepare for the next sortie. That picture is much too gentle. In remote conditions, charging is itself an operational phase with safety implications, logistical burdens and tempo consequences of its own. A battery coming back from the cold is not merely empty or partly spent. It is cold-soaked, perhaps unevenly warmed, perhaps wet with condensed moisture, perhaps handled in haste, perhaps due for inspection, perhaps already carrying more internal resistance than its outward behaviour suggests. Plugging it in is not the opposite of flying. It is another point at which the environment tests whether the programme understands its own chemistry.

That matters because remote charging is never purely electrical. Show the real scene and the fiction of simplicity falls away. A sortie ends in blowing snow or dim light. Packs are carried into a shelter or container. A generator has been running for hours. Shore power, where it exists, may vary in quality or availability. Space around the chargers is finite. Other equipment also wants the same power. Some packs are warm enough to process quickly, some are not. Logs must be updated. Cables and connectors have already lived through cold, flexing and repeated handling. Name the broader issue and it comes into focus: charging logistics. Explain it and the implication is blunt. The rate at which a site can safely generate sorties depends as much on its charging discipline and infrastructure as on the aircraft themselves.

Cold-soaked packs sit at the centre of this problem because charging is chemically less forgiving than people sometimes assume. A battery that has just come in from severe cold may still accept a connection and appear ready for replenishment. That does not mean the moment is wise. The behaviour of lithium-based chemistries during charging depends strongly on temperature, and low-temperature charging can invite degradation and, in serious cases, unsafe internal conditions. One does not need to lapse into alarmism to state the essential point. A cold battery is not simply an empty vessel waiting to be refilled. It is a stressed electrochemical system whose ability to accept charge safely and efficiently depends on its internal state, not only on its external connector. The implication for Arctic operations is immediate. Charging timing becomes part of risk management, not just battery turnaround.

This is why warming before charging is often operationally necessary even when it appears to delay everything. A team under pressure to generate sorties may look at cold packs and see lost time. A mature programme sees a queue that must be handled without damaging the very asset it depends upon. Show the decision plainly and its logic becomes clearer. The crew can begin charging immediately and preserve the illusion of tempo, or it can delay, warm and inspect, accepting a slower visible pace in exchange for battery health, safer charging and more predictable future behaviour. Name the trade-off and it is more precise than caution alone: throughput versus battery integrity. Explain it and the consequences travel far beyond that one pack. Damage introduced during hurried charging does not stay inside the charging bay. It shows up later as shorter endurance, greater voltage sag, more uncertain reserve, and a fleet that becomes less trustworthy by degrees.

Generators make this more complicated because they turn battery charging into a fuel and maintenance question as well as an electrical one. In austere sites, generators may provide the main source of power not only for chargers but for heating, lighting, tools, shelters, communications and other systems competing for the same limited output. A battery programme built around clean assumptions about charging time can therefore be upset by demands that have nothing directly to do with the aircraft. A heater for the work shelter, a communications burst, or other equipment coming online may alter what is available to the charging bench. The generator itself must be fuelled, started, sheltered, monitored and maintained. The result is that a battery charger in the Arctic is never plugged into “the grid” in the abstract. It is plugged into a chain of mechanical, fuel and operational dependencies that all carry their own failure modes. Power arrives with a supply tail.

Shore power, when available, is not automatically simpler. A harbour, station, vessel or forward facility may offer electrical supply that appears stable enough for routine charging. Sometimes it is. Yet variability in local infrastructure, extension runs, connector condition, environmental exposure and competing loads can still make charging less straightforward than its domestic counterpart. This matters because battery chargers often behave as though power quality is either adequate or not, while real operations live in the uncomfortable middle where supply exists but carries fluctuation, interruption or local improvised workarounds. A remote charging setup can therefore become a quiet source of inconsistency. The pack that charged cleanly yesterday may charge differently today not because its chemistry changed, but because the conditions of charge did. Reliability begins at the plug face.

A named example from space and planetary robotics again offers a useful mirror. Missions run by NASA and other agencies have long treated charging and energy recovery not as mundane afterthoughts but as central operational constraints shaped by environment, power source, temperature and cycle planning. The details of a Mars rover's solar management or thermal survival are obviously different from an Arctic drone pack on a charger. But the intellectual lesson is the same: energy replenishment is not a neutral reset to baseline. It is an active phase that determines what the next mission can honestly expect. Serious programmes model it, protect it and pace operations around it.

A second example comes from uncrewed and instrumented polar expeditions such as MOSAiC, where power generation and redistribution shaped what science could be performed, when equipment could be sustained, and how much operational flexibility remained once environmental conditions deteriorated. The direct relevance to drone charging lies not in identical hardware but in shared logistics. Batteries, sensors, communications systems and shelters all depend on power that must be created and prioritised in a hostile place. The lesson is not glamorous. It is severe. Energy support is itself a mission lane, and if it narrows, every downstream ambition narrows with it.

This is where sortie generation must be understood honestly. Organisations often speak as if sorties emerge from aircraft availability and weather windows. In remote Arctic conditions, charging capacity can be the real pace-setter. Not nominal charger count alone, but warmed charging positions, safe handling space, documentation discipline, connector integrity, generator fuel state, queue management and the time needed to return cold packs to a condition fit for charge. Efficiency appears here in very concrete form: whether a pack can be processed without rushing, whether chargers can be fed reliably through the night, whether crews know which batteries are warming, charging, cooling, resting or quarantined, and whether tomorrow's launch decisions begin with clear battery status rather than educated guesswork. A programme with excellent aircraft and poor charging discipline is often just a programme that has not met its bottleneck yet.

The bottleneck sharpens further when one notices that batteries do not all age alike. Packs subjected to harsher missions, poorer handling or more frequent cold exposure may begin to diverge in behaviour. This means charging is also a data problem. A site that cannot track which battery was used in what conditions, how it behaved under load, how it was handled on return, and how it responded during charging is surrendering one of the few ways it has to keep chemistry from turning into folklore. The result is a familiar form of decline. Teams begin to talk about "good packs" and "difficult packs" in anecdotal terms, not because human judgement is worthless, but because the formal battery record is too thin or too slow to carry what people have learned. Once that happens, sortie generation starts leaning on memory rather than evidence.

A fair limitation should be admitted. Not every remote charging problem is unique to the Arctic, and not every Arctic site is equally austere. Some facilities in northern Norway, Iceland, Alaska or Svalbard can provide far better support than drifting ice camps, improvised field shelters or isolated littoral outposts. Better infrastructure can soften the worst charging penalties. Purpose-built battery systems may also include management features that limit unsafe charge conditions and improve transparency about battery temperature and state. These are meaningful advantages, and they prevent the argument from curdling into melodrama.

But the counterpoint does not remove the core issue. It simply relocates it. Better sites still depend on procedures, power quality, queue discipline, storage space, thermal handling and informed judgement about when a pack is ready to charge. Battery management systems help, but they do not repeal the chemistry or the logistics surrounding it. A charger can refuse a cold pack, but the programme still needs somewhere to warm it, time to wait, people to log it, and enough other charged packs to keep operations moving without temptation. Good infrastructure does not abolish charging as a constraint. It makes the constraint more manageable for as long as the mission stays near that infrastructure.

The supply-chain character of charging becomes most obvious when something goes wrong. A generator needs attention. Fuel delivery slips. Shore power becomes unreliable. A charging connector fails. A warming shelter loses heat. Suddenly the question is not how much energy the aircraft consumes in flight, but whether the site can restore readiness before the weather window closes. This is why power planning in the Arctic belongs as much to logistics as to electrical engineering. Fuel, spare chargers, replacement leads, shelter space, warming cabinets, storage cases, inspection routines, and the mundane choreography of moving packs from one state to another all matter. The battery does not care whether the bottleneck is chemical or organisational. It simply becomes unavailable.

The quiet motif here is the socket. A socket looks like access: a point where need meets supply, where depletion ends and readiness begins. Yet in the Arctic, the socket is also a test of everything behind it. Is the power stable? Is the environment suitable? Is the pack ready? Is the queue disciplined?

Are the records honest? The simplest object in the charging chain becomes a kind of operational truth-teller. If a programme cannot answer those questions well at the socket, it is unlikely to answer them better later in the sortie.

Charging also reshapes the moral geometry of tempo. A rushed launch is obviously risky. A rushed charge is easier to romanticise because it looks like diligence, responsiveness, determination to keep the mission moving. But haste around batteries in severe cold often borrows reliability from the future. The sortie may happen. The cost appears later as degraded pack health, more uncertain cold-weather performance, a rising need for unofficial cautions, or a growing gap between displayed battery confidence and real operator trust. A well-run Arctic power system therefore has to know when not to accelerate. It must be willing to lose visible tempo in order to preserve actual capability.

That makes charging doctrine a surprisingly revealing indicator of seriousness. Where programmes are mature, batteries are not treated as generic consumables shuffled from aircraft to charger and back again. They are handled as condition-dependent assets whose readiness is shaped by recent temperature, mission profile, ageing, and the site's present power reality. Where programmes are immature, charging is treated as clerical closure to the interesting part of operations. The aircraft receives the glamour; the charging bench inherits the consequences. The Arctic is unkind to that hierarchy. It keeps reminding operators that the next flight begins, in a practical sense, not at take-off but in the discipline of the last recovery and the first safe connection afterwards.

The unresolved question is therefore harsher than it first appears. When a remote team looks at a rack of cold-returned packs and a finite source of power, what exactly is it trying to optimise: the appearance of tempo, the number of aircraft nominally available, the speed of the next launch, the long-term health of the batteries, or the safety margin that comes from knowing each pack was charged under conditions worthy of belief? In the Arctic, that choice is never merely about electricity. It is the point at which chemistry, infrastructure and supply chain either begin cooperating - or start quietly lying to one another.

Power-aware autonomy and mission rules

In January 2009, US Airways Flight 1549 lost thrust in both engines after striking birds shortly after departing LaGuardia Airport in New York. The aircraft was crewed, the context urban rather than polar, and the outcome extraordinary for many reasons. Yet one lesson belongs squarely to autonomous systems in the Arctic: energy and altitude are not abstract reserves. They are shrinking options, and the quality of decision-making depends on whether the system understands that shrinkage early enough to act usefully. Margin moves fast.

That is the proper starting point for power-aware autonomy. A battery-powered aircraft in severe cold does not simply carry a quantity of energy that ticks downward with time. It carries a condition-dependent reserve whose meaning changes with temperature, wind, mission geometry, battery age, heater load, payload demand, and the simple brutality of recovery. A machine that responds to that changing reserve intelligently can preserve itself and its mission. A machine that treats battery state as a static percentage may behave politely right up to the point where its politeness becomes dangerous. The display looks calm.

This is why autonomy tied to the energy state should be understood as a form of judgement rather than a convenience feature. Show the issue plainly, and the stakes emerge at once. A vehicle reaches a point in the route where, under benign assumptions, it can continue and still return with acceptable reserve. But the wind has shifted, the pack is colder than ideal, the payload and heaters have drawn more than planned, and the next phase includes a climb and a demanding recovery. Name what the software must do with that information and the design question sharpens: revise the mission according to live power reality. Explain why that matters, and the implication becomes obvious. An aircraft that waits until energy trouble is visible to everyone has already spent the part of the reserve that mattered most.

Dynamic return-to-home logic is one of the clearest examples of this principle. In crude systems, return triggers are fixed: at a given battery percentage or timer threshold, the aircraft abandons tasking and heads back. Fixed triggers feel reassuring because they are easy to brief and easy to display. In the Arctic, they can become a trap. A fixed threshold may be too conservative on an easy day and dangerously optimistic on a hard one. Dynamic logic improves on this by asking a better question: not "How much charge remains?" but "Given present conditions, how much energy is likely to be needed to return and recover safely?" That is a far more honest inquiry. It recognises that the value of the reserve is inseparable from the route home.

Yet the very intelligence of dynamic return logic introduces a new burden. Operators must know what the system is counting, what assumptions it is making, and what conditions will cause its estimate to change. Otherwise, prudence starts to look like caprice. Show the practical scene, and the problem becomes plain. A drone aborts a productive task leg and begins returning earlier than the crew expected. The software may be acting sensibly because wind, battery temperature or current draw has altered the recovery budget. But if the crew sees only a healthy-looking percentage and an apparently abrupt retreat, the machine's caution can feel like a malfunction. Trust becomes behavioural here. It appears as an operator who lets the aircraft execute its return without argument because the logic is legible, or as one who overrides it because the machine has become opaque at the moment it most needs to be obeyed.

Load shedding poses a similar challenge. The idea is straightforward enough. When energy tightens, non-essential functions are reduced, deferred or switched off so that flight-critical and recovery-critical functions retain priority. The logic is sound. A payload mode may be dropped, a heater cycle altered, a data link rate reduced, or a secondary sensor paused to preserve core control and navigation. Name the principle and it becomes a matter of hierarchy: not all watts deserve equal protection. Explain why this is especially important in Arctic operations and the answer is severe but simple. Cold-weather missions often drift into situations where several demands claim necessity at once. If the system cannot rank them under pressure, it is not truly power-aware. It is merely power-literate in favourable conditions.

The difficulty is that load shedding changes the mission while it is being flown. A payload operator may see image quality drop or a sensor mode disappear at precisely the moment the target becomes interesting. A supervisor may interpret reduced data transmission as a communications problem rather than a deliberate power response. A maintainer reviewing the sortie later may struggle to distinguish prudent energy management from unexplained performance inconsistency unless logs are unusually clear. This is why power-aware autonomy cannot be assessed only by whether it preserves the aircraft. It must also be judged by whether it preserves explanation. A machine that saves itself by becoming inscrutable has solved only half the problem.

A useful named example lies in planetary robotics. NASA's Mars rovers have long depended on energy-aware planning and system behaviour in which activity is constrained by battery state, thermal needs and environmental conditions. The point is not that a polar drone should imitate a rover inch by inch. The point is that serious remote systems do not treat power as a background utility. They tie it directly to mission choice, activity sequencing and survivability. Power-aware conduct is not ornamental. It is part of what makes remote operation governable when direct intervention is slow or impossible.

A second example can be drawn from the widespread use of return-to-home features in commercial and professional drone systems. These systems have made autonomy legible to a broad user base by encoding rules about low battery, lost link and automatic recovery. Their popularity is instructive precisely because it reveals both the promise and the danger of energy-linked automation. A well-understood return trigger can prevent careless overextension and normalise safe recovery. But users also know the awkward cases: automatic returns that begin at inconvenient moments, battery estimates that feel detached from wind or payload reality, and threshold behaviours that can surprise operators who know the percentage but not the underlying logic. In easy settings, that surprise may be annoying. In Arctic conditions, with longer recovery legs and less forgiveness at the end of the mission, it becomes a doctrinal issue.

Doctrine is the missing word in many discussions of power-aware autonomy. Features alone do not create safe behaviour. A programme needs shared rules about when the machine may revise mission ambition on its own, when humans may override it, and what signs indicate that an override would be reckless rather than corrective. Show the operational problem and its institutional character becomes plain. A system is configured to return dynamically based on estimated energy-to-home, but operators under pressure to complete collection begin routinely cancelling early returns because the aircraft "still looks fine". Very quickly the nominal feature and the real doctrine diverge. The software becomes a nag rather than an authority, and the programme begins teaching itself that formal energy logic is optional. The implication is stark. A wise machine inside an unwise doctrine is not enough.

This is especially true in the Arctic because energy state is entangled with factors that are difficult to perceive directly in the moment. Battery temperature is not intuitively visible. Ageing is slow until it becomes decisive. Wind aloft may differ from what the launch line suggested. Heater duty may consume more than operators sense. A route home that looked energetically easy outbound may become harder inbound when visibility, icing concerns or recovery geometry shift. That is why autonomy must translate energy state into actions and messages that correspond to operational reality, not merely numerical convenience. Percentages alone are rarely adequate. The system should help people think in terms of recoverability, not simply remaining charge.

A mature power-aware system therefore does several things at once. It estimates not only remaining energy but the cost of completing likely next actions. It compares mission continuation against recovery need. It prioritises loads according to a clearly designed hierarchy. It records why a threshold changed. It warns before it compels. It distinguishes soft caution from hard limit. None of this is glamorous. Much of it is administrative in spirit. Yet this is what turns autonomy from a promise of hands-off operation into a disciplined partnership between software judgement and operator doctrine.

Still, a fair limitation belongs here. Energy estimation is not omniscience. Models can be wrong. Conditions can change faster than the aircraft updates its assumptions. A battery that has aged unevenly or warmed unpredictably may not behave exactly as the management logic expects. Overly conservative rules can waste mission opportunity, while aggressive rules can produce self-confident failure. It would be dishonest to suggest that power-aware autonomy can eliminate uncertainty simply by calculating more elegantly.

But the limitation does not weaken the case for these features. It changes the standard by which they should be judged. The right question is not whether the system predicts perfectly. It is whether it behaves intelligibly and conservatively enough when imperfect knowledge meets shrinking energy. A machine that revises its plan early, communicates clearly, and preserves a buffer for recovery may sacrifice some efficiency in the narrow sense. A machine that waits for certainty before acting may preserve appearance while spending the very margin it was supposed to protect. The Arctic usually rewards the first kind of imperfection.

This is where operator doctrine becomes inseparable from software design. People must know what the aircraft values when energy is tight. Does it privilege immediate recovery over payload completion? Does it preserve heating over data transmission? Does it climb for safer return geometry even if that costs power now? Does it shed sensors before communications, or communications before sensing? These are not merely engineering choices. They are operational priorities turned into behaviour. If crews do not understand them, autonomy under stress becomes a source of surprise rather than a guard against it. Surprise is costly. Especially late.

The quiet motif here is the fuel gauge, though in battery systems it is more treacherous than its analogue forebear. A fuel gauge invites the belief that reserve is a quantity. Power-aware autonomy, if done honestly, reminds us that reserve is a decision space. It includes what remains, what can still be delivered, how much will be needed to return, which functions matter most, and what uncertainty should cost. A good system makes that decision space visible. A weak one reduces it to a percentage and hopes for the best.

That difference becomes sharpest near the end of missions, when every watt carries more consequence. A dynamic return trigger that starts early enough may feel irritating, especially if the aircraft seems otherwise healthy. A load-shedding event that dims a payload feed may feel like the machine is betraying the mission. Yet those feelings are not reliable measures of system quality. The better measure is whether the aircraft, its logs and its operators all remain aligned about what happened and why. Did the machine preserve recovery margin? Did it announce the reason in terms people could act on? Did doctrine make room for that behaviour, or quietly punish it?

Power-aware autonomy is therefore not best judged by elegance of interface or by the mere presence of battery features on a specification sheet. It should be judged by transparency under stress. Does the aircraft expose the assumptions behind its return logic? Does it show which loads may be shed and in what order? Does it make clear the difference between advisory and mandatory thresholds? Can a trained operator explain, before launch, how the machine will behave if cold, wind or battery age begins shrinking the energy budget faster than planned? If not, the autonomy is not yet ready for Arctic seriousness.

The unresolved test arrives at the exact moment everyone wishes it would not: when the mission still looks recoverable, the target still feels worth one more pass, and the machine begins withdrawing permission because its energy model says the safe future is already getting smaller. At that point, what matters more - the remaining percentage on the screen, the operator's intuition, or the doctrine that decided in advance how much surprise the aircraft is allowed to absorb before it must start bringing itself home?

Chapter 4

Sensors and Perception in Snow and Ice

A vehicle can carry excellent sensors into the Arctic and still come back half blind. That is the paradox. The terrain is open, the horizon broad, the targets often exposed against backgrounds that look uncluttered to the human imagination, and yet perception becomes harder rather than easier. Snow flattens contrast. Ice repeats texture. Low sun, drifting crystals and featureless light turn ordinary seeing into uncertain interpretation. The image arrives. Meaning lags behind.

The danger begins when clean imagery is mistaken for dependable perception. A video feed can look calm while withholding the depth, separation and edge definition on which real judgement depends. An infrared view can seem sophisticated while thermal contrast quietly collapses. A radar return can appear reassuringly technical while precipitation and surface clutter complicate what, exactly, is being measured. The machine sees something. That is not enough. A picture can mislead.

This matters because unmanned systems live on delegated judgement. Once operators accept that a camera, a thermal imager, a lidar unit or a fused display stands in for their own presence, the question is no longer whether data exists. It is whether the data deserves authority. Arctic conditions sharpen that question until it becomes operational rather than philosophical. A sensor that performs beautifully over temperate ground may lose discrimination over cold water and sea ice. A platform that navigates confidently with abundant visual features may become hesitant or falsely certain when the world narrows into whites, greys and weak edges. The failure is rarely theatrical. More often, it is a small drift from what the system believes it knows to what it can still justify.

That drift creates the central tension here. Everyone wants perception systems to be richer, broader and more autonomous. More sensing promises safer flight, better targeting, stronger navigation and less burden on remote crews. Yet every added layer of perception also introduces fresh dependencies: on calibration, on weather, on thermal stability, on power, on integration, and on the hidden assumption that different sensors will fail differently rather than together. The Arctic tests that assumption very hard. When the environment strips contrast, cools backgrounds towards one another, throws snow into the beam path and shakes mounting geometry through wind, vibration or ice accretion, the system may lose not one sense but several at once. Precision can become brittle.

That is why the stakes are larger than image quality or detection range taken in isolation. Perception shapes behaviour. If a sensor suite overstates certainty, operators push farther, descend lower, tighten recovery plans and trust automation longer than they should. If it understates certainty, missions grow timid, endurance is wasted and crews stop believing the platform can convert data into useful action. Trust, in this setting, is visible. It appears when a supervisor accepts a machine-generated track without a second pass, when a pilot lets an autonomous recovery continue because the cues seem credible, or when a maintenance team signs off a calibrated payload without fearing that the cold has quietly shifted its meaning. The wrong kind of confidence is expensive.

A useful way to think about the problem is to picture a windscreen. A windscreen can be technically intact and still poor at its job if it fogs, ices, distorts or catches the wrong light. The observer is not fully blind, merely compromised in ways that become serious at speed. Arctic sensing works much the same way. Visible cameras, thermal payloads, radar, lidar, inertial aids and mapped terrain each contribute a pane of glass. Some are clearer in one condition than another. Some interfere with one another's promise by drawing more power, adding weight, demanding space or creating new maintenance points. Some fail gently. Some fail with polished confidence. The challenge is not to find a perfect pane. It is to know what the whole cockpit loses when one or more begins to lie.

The path forward through this chapter follows that loss of certainty as it moves across the sensing stack. It begins with ordinary optics, where low contrast and white terrain expose how easily cameras flatter operators with crisp-looking ambiguity. It then shifts into thermal sensing, where cold backgrounds and atmospheric effects alter what heat can still reveal, before turning to active sensors whose beam paths and returns are complicated by snow, clutter and competing demands on size and power. From there, the argument tightens around fusion, calibration and drift, where the question stops being what one instrument sees and becomes whether the system can admit, contain and report degradation honestly enough to remain useful.

That honesty matters because degraded modes are not evidence of failure in the pejorative sense. They are evidence that a system has been designed to survive the disappearance of easy certainty. A platform that continues operating with reduced precision, clearer caveats and narrower authority may be far more valuable than one that preserves the appearance of exactness until the environment exposes its overconfidence. The same applies to data quality itself. Data is not merely collected; it is consumed, weighted, trusted, doubted and sometimes contaminated by conditions the system only partly understands. Once data is treated as an operational variable rather than a neutral by-product, perception stops sounding like magic and starts sounding like discipline.

The Arctic, after all, does not ask sensors whether they are advanced. It asks whether they remain interpretable when the scene ceases to offer edges for free, when thermal difference thins out, when returns are cluttered, and when calibration carries the memory of cold, vibration and ice. The sharpest question is therefore not how many sensors a platform carries, but whether their outputs still deserve belief when they begin to agree for the wrong reasons.

Optical payload limits in low contrast

On 28 November 1979, Air New Zealand Flight 901 flew into Mount Erebus on Ross Island, Antarctica. The accident had more than one cause, but one of its most enduring lessons was visual. In polar conditions, a scene can remain visible while ceasing to be readable. Snow, cloud and light can merge into a field so smooth that depth, slope and closure rate begin to lie. The eye keeps searching for an edge.

That lesson belongs not only to pilots peering through a cockpit windscreen. It belongs equally to the visible-spectrum cameras carried by unmanned systems in snow and ice. A modern optical payload can produce crisp imagery, stable framing and impressive zoom while still failing at the thing operators often assume it does best: turning a white landscape into dependable knowledge. The problem is not simply that the image looks poor. More dangerous are the moments when the image looks good enough to deserve confidence and does not.

This distinction is the beginning of any serious discussion of low-contrast perception. Optical payloads do not fail only in darkness, blizzard or obvious obscuration. They also fail in broad daylight when the Arctic scene offers too little tonal separation for the camera to preserve meaningful structure. Show the problem plainly and it becomes familiar. A drone looks down on snow-covered ground in flat light. The feed appears clean, the horizon calm, the picture technically sharp. Yet ridges, drifts, depressions, tracks and obstacles barely separate from one another because the light is diffuse, shadows are weak, and the surface reflects within a narrow range of values. Name the condition, and the engineering question sharpens: contrast starvation. Explain it, and the implication is immediate. The payload has not stopped seeing. It has stopped discriminating reliably.

Visible-spectrum cameras depend on reflected light and therefore inherit its ambiguities. In temperate terrain, cameras are often helped by shadow, colour variation, vegetation, built structures and cluttered surfaces that create edges almost for free. The Arctic withdraws those conveniences. Snow reduces texture. Overcast skies flatten illumination. Low sun angles can exaggerate or erase relief depending on orientation. Sea ice, drifted surfaces and wind-hardened snowfields repeat pattern without yielding easy scale. A human observer can sometimes compensate by bringing context, memory and suspicion to the image. A remote payload operator, staring at a stabilised video feed framed by instrumentation, can be more vulnerable than expected to the false comfort of visual neatness.

Exposure sits at the centre of this discomfort because cameras cannot record everything equally. A sensor and its processing chain must choose what part of the luminance range to privilege. Show a scene that contains bright snow, darker structural elements, weak shadow and a pale sky, and every exposure decision becomes a compromise. Expose for the snow, and darker details may sink toward insignificance. Expose for the darker subject, and snow fields may wash toward featureless brightness. Name the issue and it becomes more precise than "camera quality": dynamic range management. Explain it and the practical implication becomes harder to ignore. Detection is not only about whether an object is present in the frame. It is about whether the exposure choice preserves enough local difference for the operator or algorithm to distinguish object from background with warranted confidence.

This is where many public conversations about drone sensing become misleading. People speak as though resolution were the master variable. Higher resolution helps, of course. A better lens, a more stable gimbal, a cleaner sensor and sharper processing all matter. But sharpness cannot restore contrast that the scene is not giving. A beautifully resolved white surface can remain ambiguously white. More pixels may refine the boundaries of an uncertainty without abolishing the uncertainty itself. That is why flat light so often defeats intuition. The image looks detailed and still says very little about relief, distance or separability. Precision becomes cosmetic.

A useful historical anchor lies in Antarctic and polar flying practice, where flat light and whiteout have long been treated as more than meteorological inconvenience. They are operationally recognised because the visual world can remain present yet lose the cues needed for safe judgement. The relevance to optical payloads is direct. A camera is not exempt from the same loss of edge and depth just because it is mounted on a gimbal rather than in a cockpit. Indeed, in some respects it is more exposed to the problem because the operator watching its output cannot lean, refocus or use bodily sense in the way a person physically present might. The feed becomes the world, and that makes its omissions more dangerous.

A second example can be found in the routine experience of winter aerial photography, survey work and search operations over snow, where visible imagery often looks deceptively clear while offering poor target separability. Publicly discussed cold-weather drone use by research and inspection teams repeatedly reveals this pattern. Tracks vanish into the general surface. Small objects are missed until the camera angle changes. Surface depressions appear later than they should because weak shadowing hides their geometry. The significance of these examples lies not in any one spectacular failure. It lies in their repetition. Low-contrast difficulty is not a freak event. It is a structural property of visible imaging over snow and ice.

Angle of view matters more than many assume. A nadir-looking payload, pointed straight down, may eliminate horizon confusion while also removing oblique cues that help humans read texture and relief. An oblique view can restore some depth cues while introducing glare, haze or compressed scale. The camera does not merely record the scene; it chooses which geometry of scene perception the operator will inhabit. Show a shallow-angle view across wind-scoured snow and suddenly texture may emerge through raking light. Rotate slightly or fly under a different cloud layer and the same area becomes bland again. Name the issue and it becomes a matter of operational geometry. Explain it, and the implication is practical. Detection success may depend less on the raw capability of the payload than on whether the mission allows the right viewing angle at the right time under the right illumination.

Automatic camera behaviour introduces another layer of trouble. Many optical payloads use auto-exposure, gain control, white balance and image enhancement to keep the feed subjectively pleasing. These tools are often useful. They are also capable of concealing environmental difficulty by smoothing it into watchable video. A camera may compensate for brightness so effectively that operators underestimate how little structural information remains in the underlying scene. An image can be clean, balanced and stable while its local contrasts are too weak for dependable interpretation. That is false confidence in one of its purest forms. The display reassures exactly when caution should be rising.

The relation between image quality and trust deserves more precision than it usually receives. Trust in an optical payload is not the belief that the camera is functioning. It is the decision to act on what the image appears to show. That decision becomes visible in behaviour. An operator commits to a closer pass because the surface looks obstacle-free. A supervisor accepts a target identification without requesting a confirming angle. A vehicle continues descending because the landing zone appears legible. A planning team counts on visible imaging for a route segment because prior sorties returned aesthetically convincing footage. Once trust is described in those terms, the stakes sharpen. A payload that encourages unwarranted action is not simply limited. It is operationally hazardous.

Dynamic range is especially unforgiving in mixed scenes. Snow rarely presents itself as uniformly white under perfectly cooperative light. There may be patches of exposed rock, dark hardware, open water, vessel surfaces, or human structures that pull the camera's exposure decisions away from the subtler distinctions operators actually need. Show a drone surveying a vessel against snow-covered shoreline or sea ice and the problem becomes clear. The camera may preserve the vessel's edges cleanly while losing faint detail in the surrounding surface. Or it may expose to retain the broader bright scene and leave darker elements less informative. This is not a flaw in one particular camera. It is the unavoidable fact that a visible sensor with finite dynamic range must rank visual priorities. The Arctic often makes those priorities conflict.

A fair counterpoint belongs here. Optical payloads are not useless in snow and ice. Far from it. Under good illumination, with adequate sun angle, modest shadow, sensible exposure tuning and experienced interpretation, visible cameras can be extremely valuable. They remain unmatched for many forms of contextual understanding: recognising structures, reading markings, observing human activity, documenting surface conditions and providing intuitive scene awareness. Colour information, even muted Arctic colour information, can still matter. There are many tasks for which a visible camera remains the right primary sensor, and it would be lazy to suggest otherwise.

But the counterpoint only strengthens the larger argument. Because optical payloads can work so well some of the time, organisations are tempted to treat their failures as unfortunate exceptions rather than predictable limits. That is the error. The honest design stance is not to distrust cameras always. It is to know when the visual scene is likely to become ambiguous even before the aircraft launches. A camera that performs brilliantly in one light regime may become dangerously flattering in another. The change is not betrayal. It is physics meeting expectation.

This is why perception failure should be engineered for rather than treated as surprise. Good systems do not merely record imagery; they also reveal something about the conditions under which that imagery deserves confidence. They may expose histogram behaviour, contrast warnings, operator cues about sun geometry, or mission rules that restrict certain tasks in known low-contrast conditions. They may demand confirmatory passes from different angles, pair visible imaging with other sensing modes, or prevent automated interpretations from presenting themselves with unwarranted certainty. The point is not to make operators timid. It is to stop crisp pictures from masquerading as complete knowledge.

One of the hardest aspects of low-contrast optical work is that human beings naturally over-read smooth images. The visual system is skilled at imposing meaning on incomplete material. A faint line becomes a track. A soft tonal change becomes a ridge. A featureless patch becomes level ground simply because nothing in the image actively refutes that reading. Algorithms can be guilty of a parallel vice, committing to edge maps or object detections that depend on contrasts too weak and unstable to be trusted across frames. In both cases, the danger is not mere inaccuracy. It is premature closure: the feeling that interpretation is finished when it has barely begun.

This is where calibration and maintenance begin to matter even for visible payloads that appear mechanically healthy. A slight haze on the lens cover, minor drift in focus, small gimbal misalignment, or compression artefacts introduced in transmission may all shave away what little contrast the scene still offers. In benign settings such losses may be tolerable. In snow and ice they can become decisive because the optical budget is already thin. The operator may blame the environment entirely and miss the fact that the payload has been contributing its own small degradations. Low contrast punishes every weakness in the chain.

The quiet motif in all this is the edge. Operators search for it instinctively: the boundary of a snowdrift, the line of a track, the profile of a ridge, the silhouette of a person or structure against a pale field. Edges are what turn sight into interpretation. Arctic optical operations become difficult not because light disappears, but because edges do. Or, perhaps worse, because they appear just strongly enough to invite confidence without quite deserving it. The job of engineering is therefore not only to improve the image. It is to ask when the edge has become too weak to carry decisions safely.

That question reaches beyond the payload itself into doctrine. If a route plan assumes that a visible camera can verify surface conditions before recovery, what happens when flat light makes verification ambiguous? If a search task depends on detecting small objects against snow, what level of contrast loss forces a second sensor or a different geometry? If an operator says "I think I see it", what standards of confirmation exist before the aircraft commits fuel, time or risk on that basis? These are not merely tactical choices. They are perception rules shaped by respect for known optical limits.

The unresolved tension, then, is not whether visible-spectrum cameras belong in Arctic operations. They do. It is whether the system around them is honest enough to distinguish a beautiful image from a trustworthy one. When the feed looks clean, the horizon calm and the snowfield deceptively legible, what exactly tells the operator that the camera is still showing enough - and not merely showing something pleasingly white where the missing edge used to be?

Infrared performance and atmospheric effects

In February 1991, during the Gulf War, television audiences were introduced to a certain myth of thermal vision: the idea that infrared sensors could peel back darkness and confusion alike, revealing the world with almost indecent clarity. Heat seemed to promise a kind of honesty that ordinary sight could not. The image glowed; the target emerged; uncertainty appeared to retreat. That promise has lingered ever since, and it travels easily into discussions of Arctic autonomy. If visible light struggles over snow and ice, surely heat will rescue perception.

It will not, at least not so neatly. Infrared sensors are powerful because they do not depend on reflected visible light in the ordinary way, but they are not exempt from physics any more than cameras are. They need difference. A thermal imager sees best when something in the scene emits, retains, sheds or reflects heat in a way that stands apart from its background. The harder truth is that Arctic environments often reduce that difference rather than heighten it. The world does not go dark. It goes thermally quiet.

That quiet is the starting point for serious infrared work in snow and ice. Show the scene plainly and the problem becomes clear. A sensor that distinguishes vehicles, people or infrastructure over temperate ground may be pointed over cold ocean, sea ice or snow-covered land where surfaces, air and objects have drifted towards a narrower thermal band. Name the governing concept and the engineering challenge sharpens: thermal contrast. Explain it and one sees why so many expectations misfire. Infrared performance does not depend only on how warm a target is in absolute terms. It depends on how separable the target is from what surrounds it, how that difference survives the atmosphere between sensor and scene, and how stable it remains over time. The implication is severe. A target can still be present, still be physically meaningful, and yet cease to stand out enough for reliable detection or interpretation.

This is why the Arctic often disappoints those who treat infrared as a universal corrective to visible-light limitations. Snow and ice can look visually featureless, which tempts people to assume they will offer a clean thermal backdrop. Often they do not. Cold ocean surfaces, sea ice, packed snow, exposed rock, metal, wet surfaces and wind-cooled equipment can all interact in ways that flatten expected differences. Wind matters especially because it strips heat from exposed objects, reducing the contrast between a target and its environment. A piece of equipment or a vehicle that might present a clear thermal signature after sitting in calm air may become far less distinct after sustained wind has drawn its surface temperature down towards ambient conditions. The image remains crisp. The target softens into the field.

A simple example captures the point. Consider a small craft or item of infrastructure over cold water or sea ice. In temperate conditions, one might expect the object to read clearly against the background because it has retained heat differently from the ground or sea. In Arctic conditions, especially after prolonged exposure, wind and low ambient temperature can strip away much of that differential. The sensor still reports a scene. But target separability declines because everything has moved closer together thermally. That phrase deserves respect. Reduced separability is not a minor inconvenience. It is the difference between detection that can support action and detection that only flatters retrospective interpretation.

Background temperature is therefore not a passive setting. It is part of the contest. A sensor may perform impressively over land warmed unevenly by sunlight, vegetation and built surfaces, where thermal textures are abundant. Transfer the same sensor to a snowfield, an ice edge or a maritime Arctic scene and the background may cease to provide the same helpful variability. The result can be oddly deceptive. A system with strong technical specifications and good results elsewhere appears to "underperform" in the Arctic, when in truth it is encountering a scene that simply offers less thermal structure for the same technology to exploit. Performance did not vanish. The environment changed the meaning of good enough.

Atmospheric effects deepen the difficulty. Infrared sensing relies on radiation travelling from target to detector through air that is rarely inert. Snowfall, blowing crystals, moisture, mist, low cloud, temperature gradients near the surface, and maritime air over cold water can all alter what survives the journey. Show the operational problem and it becomes familiar. A drone flying over open leads, broken ice or cold littoral water returns a thermal picture that seems plausible at first glance. Yet atmospheric moisture and surface-layer effects have already softened edges, attenuated contrast or introduced a haze of thermal ambiguity that the operator may not fully appreciate. Name the phenomenon and it becomes more than "bad weather": propagation loss and scene contamination. Explain it and the implication follows. Infrared does not fail only when the target is cold. It also fails when the air between sensor and target begins to behave like part of the image.

This matters because many users experience infrared through polished displays that encourage confidence. A thermal image looks technical, often monochrome or false-colour, stripped of the familiar clutter of daylight imagery. That very austerity can mislead. Operators may assume that because the image seems less noisy, it is more truthful. In Arctic operations the reverse can be true. The thermal image may be aesthetically simpler because much of the scene has converged towards sameness. What is missing is not mess but difference. The display has cleaned away some of the uncertainty by rendering it as smoothness. False confidence begins there.

A named historical example helps locate the larger issue. During the Falklands War in 1982 and in later cold-weather military development, infrared and night-observation systems were increasingly valued for their ability to offset darkness and poor visual conditions. Yet military and aviation practitioners alike learned that environmental conditions could transform performance in ways that no brochure could abolish. Cold, moisture, wind and background effects mattered. The enduring lesson is not that infrared is unreliable. It is that thermal sensing must always be read against the environment that produces the thermal scene, not merely against the sensor's nominal capabilities. Arctic operations make that lesson less optional and more brutal.

A second example appears across search, inspection and wildlife observation work in northern and maritime environments, where thermal imagers can be remarkably useful one hour and markedly disappointing the next. Publicly described operations often note that detection success depends not simply on target type but on time of day, recent exposure, wind, surface conditions and whether the target has retained or shed heat differently from its surroundings. A person, animal, vessel component or structure may show clearly after one thermal history and much less clearly after another. The point is not that field operators are fickle. It is that the thermal scene is historical. What the sensor shows now depends on what target and background have been doing for some time.

This historical quality of thermal imagery is often overlooked. Visible images tempt us to think spatially: what is where. Infrared demands more temporal thinking: what has warmed, cooled, retained, conducted, evaporated, frozen, or been stripped by wind. A metal object exposed to sun may remain visible after light fades. The same object after prolonged cold wind may nearly disappear into its background. Fresh snow may insulate one surface while exposing another. A vehicle recently operated may glow in ways that a long-idle one will not. The sensor sees a thermal past as much as a present. That is both its power and its trap.

Sensor settings further complicate interpretation. Polarity choices, contrast stretching, gain control and image enhancement can help operators see weak differences, but they can also exaggerate minor thermal variation into apparent significance. Show a scene with narrow temperature spread and the processing chain may still produce a visually dramatic image by stretching what little difference exists. The result can look rich while remaining fragile. A faint contrast becomes bold on screen, and with that boldness comes temptation: perhaps the target really is there, perhaps the edge really is meaningful, perhaps the apparent hotspot deserves action. Yet what the image has amplified may be marginal enough to vanish with a slight change in angle, range or wind. Processing can clarify. It can also seduce.

This is where operator doctrine and engineering discipline must meet. Infrared imagery should not be judged solely by whether something warm appears. It must be judged by whether the system can report, or the crew can infer, how trustworthy that appearance is under current conditions. Trust again becomes concrete. It appears when a supervisor accepts a thermal contact as sufficient for route adjustment, when a crew commits to a closer inspection based on a weak hotspot, when an operator discounts a non-detection because wind and background have probably flattened contrast, or when a team delays action until another angle or sensing mode confirms what the thermal view only hints at. These are operational decisions, not abstract reflections on "confidence".

A fair limitation deserves emphasis. Infrared can be exceptionally valuable in the Arctic. Against the right background, at the right time, with a target whose thermal history favours separation, thermal imaging can outperform visible-spectrum cameras decisively. It may detect living beings, machinery, exhaust patterns, structural anomalies or ice features that ordinary optics miss altogether. Darkness is not the only case where it helps. Low visible contrast does not automatically mean low thermal contrast, and many missions would be poorer or riskier without infrared capability. Any serious account must grant this.

But the counterpoint only sharpens the central argument. Because infrared can be so effective in certain Arctic conditions, operators and institutions are tempted to generalise that success across conditions where it does not belong. They speak of thermal sensing as though it were a layer of certainty laid over the world, when in fact it is another mode of interpretation with its own dependencies and blind spots. The most costly errors tend to arise not from using infrared, but from assuming it is seeing more than the thermal environment presently allows.

This is why atmospheric humility matters. Snow in the air, blowing crystals, maritime moisture, cold-surface inversions and long oblique sightlines can all degrade performance in ways that are not obvious from the image alone. A thermal camera may continue to produce stable video and plausible contours while the atmosphere has already thinned the distinction between meaningful signal and background confusion. If operators are not trained to recognise those circumstances, or if systems do not help expose them, the sensor becomes less a tool than a polished guess.

The quiet motif here is the glow. People are drawn to it because it feels like revelation. Something emerges from darkness or sameness and seems to announce itself. But a glow is only useful if it represents genuine separability rather than processing, wishful interpretation or an environmental moment that will not survive the next change in wind or angle. Arctic thermal operations are difficult not because the glow disappears entirely, but because it can remain just strong enough to invite belief when the background is already catching up.

That is why thermal sensing in the Arctic should be read as a contest of temperatures, histories and atmospheres rather than a simple contest of sensor quality. A better imager can help. So can careful mission timing, shorter slant ranges, awareness of wind effects, and sober understanding of how cold water, ice and snow flatten scenes. Yet none of these abolishes the governing question. When the target, the background and the air between them all begin approaching one another thermally, what exactly is the sensor still telling the truth about - and what is it merely rendering in a persuasive monochrome?

Radar and lidar in precipitation and clutter

In February 1943, during the Battle of the Bismarck Sea, radar helped aircraft and ships fight through darkness and weather in ways that had seemed improbable only a few years earlier. Since then, active sensing has acquired a certain aura of emancipation: if vision is poor, emit energy, read the return, and let the world disclose itself. That promise has travelled easily into Arctic operations. When snow flattens optics and thermal contrast weakens, radar and lidar are often imagined as the next clean answer.

They are not clean answers. They are powerful ones, and sometimes indispensable, but they arrive with their own bargain. Active sensors do not merely observe a scene; they interrogate it. The scene answers back through precipitation, surface texture, motion, multipath effects, beam geometry and the hard limits of platform power and payload space. In snow and ice, those answers can be noisier and more ambiguous than the elegance of the hardware suggests. The beam goes out. Confusion returns.

That is the starting point for serious use of radar and lidar in the Arctic. Show the situation plainly and the problem becomes concrete. A vehicle flying over sea ice or a snow-covered littoral in degraded visibility turns to active sensing because visible imagery has become thin and thermal separability uncertain. The sensor begins returning data. Yet some of that data now comes not from the object or terrain of interest but from the intervening snow, from blowing ice crystals, from the rough surface itself, or from repeated reflections off awkward geometry. Name the governing issue and the discussion becomes more precise than "bad weather": clutter and attenuation. Explain it, and the implication is immediate. Active sensing does not abolish the environment. It measures through it, against it and sometimes partly because of it.

Radar and lidar differ in method, but they share this burden. Radar uses radio waves and can often reach farther and penetrate conditions that would trouble optical sensors. Lidar uses laser light and can provide fine spatial structure, sometimes with striking geometric clarity. Both therefore tempt designers into a familiar conclusion: combine them with other sensors, and perception will finally become dependable. Yet the Arctic punishes that conclusion when it is held too casually. Snow and suspended crystals can scatter lidar returns, filling the system with spurious near-field points or degrading confidence at the very ranges where obstacle detection matters most. Radar, though generally more tolerant of obscurants, must still contend with sea clutter, ice edges, ridgelines, multipath reflections and the fact that rough, cold environments can produce returns rich in energy but poor in easy interpretation.

Lidar is especially vulnerable to the seduction of visual neatness. A lidar point cloud or derived surface model can look wonderfully exact. Show a clean, structured return from a benign setting and it feels like geometry itself has become available to the aircraft. Move that same system into blowing snow or drifting crystals, however, and the point cloud can begin to populate with atmospheric debris masquerading as spatial fact. The sensor is not "broken". It is faithfully reporting backscatter from material now present in the beam path. Name the problem and its character becomes clear: particulate contamination of the return. Explain it, and the operational consequence follows. A platform relying heavily on lidar for low-altitude obstacle awareness or landing assessment may become more conservative, more confused or falsely busy precisely when the operator hoped active sensing would simplify the scene.

Radar suffers a different form of disappointment. It is often praised, rightly, for seeing through darkness and many forms of obscuration better than cameras can. But Arctic surfaces are not passive backgrounds waiting to be ignored. Sea ice, broken water, rough snow, pressure ridges, vessel superstructures and coastal geometry can all generate strong returns. Some are useful. Some are simply loud. Show a radar picture over a cluttered icy littoral and the engineering challenge becomes obvious. The sensor has signal. Perhaps too much of it. Name the condition and the task sharpens: discrimination under clutter. Explain it, and the implication is serious. A strong return does not equal a useful return if the system cannot distinguish target from background structure in a way that supports real decision-making.

The maritime world offers a long history of this lesson. Marine radar has saved ships for decades, but experienced navigators have also known that sea clutter, weather echoes and awkward coastal reflections complicate interpretation even in comparatively familiar waters. Move into Arctic conditions, where ice edges, mixed sea states, variable salinity effects, and irregular frozen surfaces add new texture to the return, and the interpretive burden rises again. The sensor remains valuable. The scene becomes harder. That is the pattern worth preserving from maritime experience: active sensing extends perception, but it does not simplify the world into innocence.

A second, more recent example comes from automotive lidar and radar development for driver assistance and autonomy, much of it publicly discussed in relation to snow, rain and low-visibility edge cases. These systems have shown both the promise and the fragility of active sensing in adverse weather. Lidar can generate exceptionally detailed environmental models in clear conditions, then struggle when precipitation and suspended particles clutter the near field. Radar can remain functional in obscurants that degrade cameras, yet still face difficulties with ghosting, multipath and target classification in complex scenes. The relevance to Arctic drones is direct. Cold-weather autonomy does not inherit some magical version of active sensing exempt from the burdens already visible in other domains. It inherits those burdens and adds harsher geometry, stricter power limits and more difficult recovery consequences.

Range, resolution and power draw sit at the centre of the design bargain. This trio should be treated not as a menu of independent virtues but as a linked constraint. Show the practical design problem and the shape of the trade-off becomes plain. A longer-range radar may demand more power, larger antennas, different mounting compromises or more processing. A finer-resolution lidar may consume power and data bandwidth while also narrowing tolerance for contamination, icing or alignment drift. A small platform that carries an active sensor does not merely gain perception.

It inherits energy demand, thermal burden, integration complexity and sometimes a new vulnerability to misalignment or surface fouling. The implication is one the Arctic keeps forcing back into view. A new sensor is never only a new capability. It is also a new maintenance and power problem.

This matters because platform sizing is shaped by sensing choices more than brochure language usually admits. A radar set with useful range and processing demands may require a larger airframe, stronger power margins, more thermal management and more conservative endurance planning than a lighter passive payload. A lidar added to support precise landing or terrain following may seem modest until its power draw, cooling needs, protective window, mounting stability and data handling chain are fully accounted for. The aircraft then changes around the sensor. It may need a bigger battery, which changes mass, which changes endurance, which changes recovery margin, which changes mission design. One sensor addition can ripple through the whole vehicle. That is why integration burden is not a side issue. It is often the real issue.

Icing and contamination deepen this burden further. Active sensors do not only suffer from what fills the air beyond the platform; they also suffer from what accumulates on the platform itself. A radar radome or lidar window carrying ice, frost, salt residue or packed snow has already changed the outgoing and incoming path before the beam reaches the environment. Show the field reality and the risk becomes practical rather than theoretical. The aircraft returns from a sortie, is brushed down, perhaps inspected quickly, then launched again. A faint film remains, or a window heats unevenly, or a protective cover carries micro-abrasion from repeated exposure. The sensor still operates, but its performance has shifted. Operators may blame the weather entirely and miss that the platform has begun contributing some of its own degraded sensing. Active perception is never just out there. Part of it is on the nose of the aircraft.

This is one reason active sensing can generate false confidence of a distinctive kind. Optical or thermal weakness often looks like weakness. Radar and lidar weakness may arrive as data richness. The screen fills. The point cloud thickens. Returns abound. Yet abundance is not clarity. A cluttered radar return, a near-field snow halo in lidar, or a neatly rendered but partially contaminated landing surface can all encourage the feeling that the machine is seeing a great deal when in fact it is struggling to sort relevance from intrusion. Trust therefore has to be described behaviourally. It appears when operators accept a terrain-following estimate, a landing cue or an obstacle warning without unnecessary hesitation. It also disappears in concrete ways: extra orbits before descent, rejected automated landings, wider obstacle margins, more frequent overrides, or an institutional habit of treating certain weather as "technically flyable but not really trustworthy".

A fair limitation should be granted clearly. Radar and lidar can be extremely useful in Arctic operations, and either may outperform visible and thermal sensors decisively under certain conditions. Radar may preserve situational awareness where snow and darkness cripple ordinary imaging. Lidar may supply precise local geometry during landing, inspection or navigation tasks when contrast in the visible scene is too weak to rely on. Some platforms and missions can tolerate the power draw, carry the payload, and maintain the sensor with enough discipline that active sensing produces genuine operational value. It would be lazy to imply otherwise.

But the counter-example only sharpens the central claim. Where active sensing works well in severe conditions, it is because the system around it has acknowledged the full cost of making it work: energy, weight, thermal control, protective surfaces, calibration, processing, maintenance and operator training. Success does not mean the sensor transcended the Arctic. It means the programme paid for an honest operating envelope. The reconciliation is therefore straightforward. Radar and lidar are not silver bullets, but neither are they luxuries. They are demanding instruments whose value depends on whether the platform can afford, physically and institutionally, the obligations they impose.

That institutional part matters more than it first seems. Operators and maintainers need a language for distinguishing sensor limitation from sensor failure. Was the landing-assist lidar misreading because of blowing snow in the beam, because the window carried residue, because the vehicle attitude induced awkward geometry, or because the underlying terrain produced confusing reflectivity? Was the radar contact weak because the target was small, because sea clutter rose, because the beam placement was poor, or because power settings were constrained by the rest of the aircraft's energy budget? If teams cannot answer such questions with reasonable confidence, the active sensor becomes another producer of mystery rather than a reducer of it.

The quiet motif here is the beam itself. It feels assertive: something sent outward to compel a reply from the world. That is part of why active sensors are so alluring in snow and ice. But the beam does not pass through a neutral theatre. It passes through moving crystals, uneven moisture, dirty windows, vibrating mounts, cluttered surfaces and a power budget that may already be under strain. What comes back is therefore not only a portrait of the scene. It is a record of the whole path. The better the operator or system understands that path, the more useful the return becomes.

This is also why active sensing and autonomy must be evaluated together. A radar or lidar that merely produces data is only half integrated. The real question is how the aircraft interprets that data under stress. Does the system know when clutter has increased enough to reduce authority? Can it distinguish likely atmospheric contamination from solid obstacles? Does it narrow its behaviour gracefully when the active sensor becomes less trustworthy, or does it continue presenting precision as though the beam had returned from a tidy world? These are not narrow technical curiosities. They define whether the sensor adds resilience or simply new ways to be wrong.

The unresolved question is therefore harder than whether radar or lidar should be carried. Many platforms will answer yes. The sharper question is whether the aircraft can afford the full truth of that answer: the added watts, the extra weight, the surface that must stay clear, the calibration that must hold, the processing that must not lag, and the operator doctrine that must know when a screen full of returns has stopped being knowledge and become merely weather answering back.

Sensor fusion and degraded-mode operation

On 1 June 2009, Air France Flight 447 disappeared over the Atlantic after unreliable airspeed data triggered a chain of confusion between sensors, automation and human interpretation. The aircraft was not operating in the Arctic, and the accident cannot be reduced to a neat lesson about any single instrument. Yet one truth from that night belongs in every harsh-environment autonomy programme: a system can possess many sources of data and still lose its grip on reality if those sources begin failing together, or if the logic combining them cannot admit uncertainty quickly enough. More sensors do not guarantee more truth.

That is the promise and the danger of sensor fusion in snow and ice. Fusion sounds like prudence made technical: combine visible imagery, infrared, radar, lidar, inertial estimates, GNSS, altimetry and map data, and each weakness will be covered by another strength. Sometimes that is exactly what happens. But Arctic conditions are unusually good at undermining the comforting assumption behind the idea, which is that different sensors fail differently and independently. In low contrast, blowing snow, icing, vibration and repeated thermal cycling, inputs can become correlated in their weakness. The system appears rich in evidence while its distinct channels are all beginning to lean in the same wrong direction. Agreement can deceive.

This is why fusion must be understood as a discipline of scepticism rather than a celebration of abundance. Show the practical case and the issue becomes clear. A drone in poor visibility combines camera imagery, inertial data, GNSS, a range sensor and perhaps some radar or lidar assistance. Each stream looks plausible enough. Yet the camera has lost contrast, the lidar is seeing suspended particles, the GNSS geometry is less helpful than expected, the inertial estimate is drifting slowly, and the range picture is being complicated by clutter or surface ambiguity. The fused output may still present a smooth estimate of where the aircraft is and what lies beneath it. Name the problem and the engineering stakes sharpen: correlated degradation. Explain it and the implication becomes harshly practical. A fused solution can become most persuasive at the moment it deserves the most doubt.

The common defence against this problem is diversity. The principle is sound. If one sensor depends on visible texture, another may depend on thermal difference; if one depends on direct line of sight, another may tolerate obscurants better; if one drifts slowly, another may provide an external correction. Diversity is what makes fusion worth attempting at all. But diversity in hardware does not automatically create diversity in environmental vulnerability. Arctic weather and surface conditions can erase visible contrast, flatten thermal contrast, clutter active sensing and stress calibration together. Even when the sensors themselves fail by different mechanisms, the operational result can converge towards the same outcome: weaker separation between target and background, poorer altitude confidence, reduced obstacle awareness, or growing ambiguity in navigation. The system is diverse on paper and brittle in weather.

This is where degraded-mode operation becomes more important than nominal precision. A degraded mode is not a confession of defeat. It is a way of continuing usefully after the environment has made exactness dishonest. Show the contrast and the point grows plain. One system keeps presenting precise-seeming position, track or landing cues even as its input confidence thins. Another withdraws gracefully into broader but more dependable behaviour: wider margins, simpler route structure, more conservative altitude keeping, reduced dependence on automated landing, slower manoeuvre, limited payload functions, clearer warnings.

Name the principle and it becomes one of the most mature ideas in harsh-environment autonomy: bounded usefulness. Explain it, and the implication follows. In Arctic operations, a machine that knows how to become less clever is often more valuable than one that remains elegant until it suddenly is not trustworthy at all.

The superiority of bounded usefulness over collapsing precision is not merely philosophical. It is visible in decisions. A vehicle in degraded mode may loiter rather than descend, return rather than continue, refuse a close inspection pass, or abandon precision recovery for a safer alternate procedure. Those actions can look disappointingly cautious if one judges them against the best-case capability of the platform. They look different if judged against the alternatives actually available in snow, ice and uncertain sensing. Trust becomes legible here. It appears when an operator accepts the machine's narrowed authority because the reduction is intelligible, or when a supervisor endorses an earlier abort because the degraded mode preserved recoverability. A system that degrades well does not ask for blind faith. It shows its caution in forms people can work with.

A useful historical example comes again from aviation automation, including the long inquiry that followed Air France Flight 447. The relevance is not that Arctic drones will replay a transatlantic airliner tragedy in miniature. It is that failures in complex systems often become more dangerous when the handover between normal mode and degraded mode is badly understood. Data does not need to disappear entirely for the relationship between machine confidence and human confidence to fracture. Unreliable or disputed sensor input can alter automation behaviour, protections and assumptions, while operators struggle to understand what has changed and why. Arctic sensor fusion faces a version of that same challenge whenever the environment erodes several inputs at once. The system must not only detect the erosion. It must also express its consequences without delay or vanity.

A second example can be found in planetary robotics, where NASA missions have long fused inertial sensing, visual odometry, terrain models and other sources to navigate in environments that are uncertain, remote and hostile to easy intervention. These programmes matter here not because Mars is a direct stand-in for the Arctic, but because they show a mature stance towards uncertainty. The most successful remote systems do not assume that all sensing will remain equally informative. They are built to retreat into safer and simpler modes when parts of the perceptual stack weaken. The lesson worth borrowing is not any single algorithm. It is the institutional humility embedded in the logic: uncertain knowledge should narrow the action set before it contaminates the whole mission.

That humility is often missing when organisations speak too casually about fusion as though it were a cure for bad weather. Fusion does not cancel uncertainty. It redistributes the labour of uncertainty across sensors, models and decision rules. That labour must be designed. A system needs to know not only how to combine inputs but how to compare their failure patterns, how to notice when apparent agreement is suspicious, and how to assign less authority to data that remains numerically stable while becoming contextually weak. This is where the engineering becomes epistemic in the strict sense. The question is no longer only what the machine sees. It is how the machine knows that it knows too little.

One of the most difficult cases arises when the fused solution remains smooth. Engineers and operators alike are easily comforted by smoothness. A trajectory estimate without jumps, a stable obstacle map, a calm landing cue or a consistent track line feels more trustworthy than one full of noise and hesitation. Yet smoothness can be a product of filtering and model assumption rather than environmental truth. In bad Arctic conditions, a fused estimate may grow more stable precisely because the system is leaning harder on priors, inertia or modelled expectation while live sensing becomes less informative. The result can be dangerous. The machine looks composed because it is slowly substituting continuity for evidence. Degraded modes exist to interrupt that temptation.

This is why fusion should be paired with explicit confidence discipline. Not generic confidence scores displayed as decorative percentages, but operational signals that change what the system is allowed to do. A drop in confidence should alter mode authority, route freedom, descent permissibility, landing logic, or reliance on specific sensor classes. Show the alternative and its weakness becomes obvious. If confidence estimates do not change behaviour, they remain informational garnish. If they do change behaviour, the system begins to protect itself against its own hunger for elegant answers. The Arctic rewards the second posture. It is hard on any design that treats uncertainty as something to be hidden behind a polished interface.

A fair limitation deserves room. Sensor fusion can genuinely improve robustness, and in many missions it already does. Combining complementary sensors often yields better navigation, better obstacle awareness, better target detection and better resilience than any one instrument could provide. It would be perverse to argue for simple single-sensor purity in environments that so often strip away individual sensor performance. Likewise, degraded modes can become too conservative if tuned poorly. A system that narrows its behaviour too early may waste opportunity, frustrate operators and reduce the value of an expensive platform. Caution can become its own kind of brittleness.

But the limitation only sharpens the main point. Fusion is most useful when it is designed with a clear account of how correlation can emerge under stress. Degraded modes are most useful when they preserve enough function to matter without pretending that the original precision remains justified. The reconciliation lies not in choosing between bold capability and timid fallback, but in deciding exactly how much authority the system should retain as evidence quality worsens. That is not a cosmetic design choice. It is the moral geometry of autonomy under uncertainty.

This geometry extends into maintenance and calibration as well. A fused system inherits the weaknesses of its components, mounts and alignments. A slight mechanical shift in a lidar unit, a camera boresight drift, a quietly accumulating inertial bias, a radar installation angle altered by icing or vibration - each can be small enough to pass unnoticed alone and large enough to distort fusion once external references degrade. The fused output may remain plausible because the software is skilled at reconciling disagreement. That reconciliation can buy time. It can also mask root causes. A serious Arctic programme therefore treats degraded mode not merely as an in-flight concept but as a design philosophy that continues through inspection, calibration and fault analysis. A machine that degrades gracefully should also be diagnosable gracefully.

The quiet motif here is the braid. A braid is stronger than a single strand because its fibres support one another. Yet if enough strands are wetted, frozen and pulled in the same direction, strength becomes more complicated than multiplication. Sensor fusion works much the same way. Its promise lies in combination. Its danger lies in forgetting that a combination can also transmit shared weakness. The braid must be inspected, not simply admired.

This is also why degraded mode must be understandable to people. A machine that quietly tightens its own constraints without making the reason legible invites override, resentment and eventual doctrinal drift. Operators need to know what has been lost, what remains reliable, and what the platform will no longer attempt. Otherwise, degraded mode appears as obstinacy rather than prudence. The best systems therefore make their retreat intelligible. They do not merely do less. They say, through behaviour and interface, why less is now the honest quantity. That honesty is part of survivability.

Usefulness under uncertainty sounds modest compared with the rhetoric often attached to autonomous sensing. It is also much harder to achieve. It asks designers to resist the seduction of perfect fused world models and to build systems that can still contribute after that perfection has become untenable. It asks operators to value recoverable ambiguity over collapsing confidence. It asks institutions to reward an aircraft for returning with narrower but truer perception rather than punishing it for abandoning precision that no longer deserved belief. These are not aesthetic preferences. They are operating principles for environments that erase easy reference.

The unresolved test comes when the sensors still agree enough to look coherent, the mission still appears feasible, and the weather has not become dramatic enough to justify obvious alarm. That is when the strongest system is not the one that squeezes one more performance claim from a weakening perceptual stack, but the one that knows exactly when to become simpler, broader and less proud. The hard question is whether the fusion logic, the operators and the mission rules all recognise that moment soon enough - before precision turns from asset to trap.

Calibration, icing, and mechanical drift

On 1 February 2003, the Space Shuttle *Columbia* broke apart during re-entry after damage sustained at launch had altered the vehicle in ways the mission could not ultimately absorb. The catastrophe had nothing to do with Arctic drones, and it would be foolish to force a false equivalence. Yet one hard engineering truth travels cleanly from that event into cold-weather autonomy: a system can continue operating for some time after a physical change has already made its internal assumptions less true. The machine keeps going. That is the danger.

Calibration lives inside that danger. People often speak of calibration as though it were a tidy, one-time act carried out on a bench by careful hands and then preserved until another formal service interval arrives. Severe environments are much less polite. In the Arctic, calibration is not merely set and checked. It is stressed, nudged, cooled, shaken, iced, warmed, handled, and occasionally lied to by the very hardware meant to preserve it. A sensor may remain fully alive while becoming slightly less honest. That slightness is exactly what makes it hard.

The central problem is not that cold weather causes obvious instrument failure every time it arrives. More often it causes small departures from alignment, bias or reference that remain hidden while other systems are strong enough to compensate. A compass may drift a little. An IMU may carry a bias that grows with temperature shift. A gimbal may sit a fraction differently on its mounts after icing and vibration. A lidar or camera may no longer point exactly where the system assumes it points. Show the sequence plainly and the issue becomes operational rather than abstract. The aircraft launches, flies apparently well, and completes most of its tasking. Only later, perhaps when GNSS quality worsens, when visual references thin out, or when automated recovery depends more heavily on internal state estimates, does the small misalignment stop being tolerable. The reference moved.

That is why calibration should be understood less as a number than as a relationship. A calibrated system is one in which the machine's internal account of orientation, position, rate and sensing geometry still corresponds closely enough to the real world to support decisions. Name that relationship properly and the argument sharpens: correspondence under stress. Explain it and the implication becomes severe. When correspondence degrades, the system may not fail immediately. It may continue operating on borrowed coherence, with other sensors and control logic quietly masking the shift until the moment masking is no longer possible. Arctic operations are full of such moments because visibility, connectivity and external references often degrade unevenly rather than all at once.

Temperature plays an early role here because many sensing systems are thermally sensitive even when their designers account for that sensitivity in principle. Electronics drift. Mechanical dimensions change. Mounting tensions alter. Adhesives and potting compounds behave differently. Magnetic characteristics can shift subtly with environment and nearby power states. A calibration established at one temperature may not hold perfectly at another, still less across repeated transitions between cold soak, powered warming, flight loads and sheltered maintenance. The practical point is not that every sensor becomes untrustworthy in the cold. It is that cold changes the burden of proof. A well-calibrated payload in a stable workshop is not the same thing as a well-calibrated payload after repeated exposure to Arctic cycling. Stability is earned again and again.

Compass behaviour illustrates the issue with unusual clarity because it sits at the intersection of simplicity and treachery. The idea of a compass sounds almost antique in an age of fused navigation and satellite positioning yet heading reference remains foundational for many systems. In Arctic operations, magnetic behaviour is already complicated by latitude, local anomalies and installation effects. Add temperature shifts, structural changes, nearby current draw, icing around mounts or cable routing strain, and a compass can begin to contribute small heading errors that are easy to miss while GNSS is healthy and harder to ignore once GNSS confidence dips. A hidden bias is not dramatic on a status page. It becomes dramatic when the aircraft begins to trust it more than it should.

IMUs pose a related but distinct problem. Inertial sensors are invaluable precisely because they remain available when external references falter. That same value makes their biases operationally dangerous. A gyroscope or accelerometer with small, temperature-sensitive errors may appear harmless while external corrections arrive frequently. Once those corrections thin out, the inertial system starts carrying more of the navigation burden, and the hidden bias becomes motion, attitude error or drift in the machine's own story about where it is. Show the operational pattern and the lesson is plain. A platform that appears stable in good conditions grows oddly tentative or overconfident in degraded ones because a bias that was once masked is now steering the estimate. The sensor did not suddenly become bad. The context stopped forgiving it.

Icing adds a more mechanical form of dishonesty. When ice accretes on sensor housings, gimbal assemblies, antenna fairings or external mounts, it does not merely block light or add weight. It can change alignment, alter resonance, create asymmetrical loading and disturb the delicate physical relationships on which calibration depends. Show a camera or lidar unit after a sortie through freezing moisture and blowing snow, and the temptation is to ask only whether the aperture or window remains clear. The harder question is whether the sensor is still pointing exactly where the software believes it is pointing. A minute angular shift may be invisible to casual inspection and still sufficient to distort georeferencing, tracking or automated cueing. Ice can therefore turn a calibration problem into a systems problem without ever appearing as a spectacular sensor failure.

Vibration does similar work more quietly. Arctic conditions impose vibration not only through flight itself but through rough transport, repeated handling, gusty recovery, hard landings, deck motion, rotor wash recirculation and the simple fact of operating over surfaces and in winds that are less forgiving than those of temperate test ranges. Vibratory exposure does not need to crack something to matter. It may loosen a mount microscopically, alter preload in a fastener, shift a connector's seating, or encourage wear in a bearing or isolator. The resulting mechanical drift can be too small for immediate detection and still large enough to move a sensor out of its assumed relationship with the airframe. This is why harsh-environment calibration should never be treated as an abstract software property. It lives in screws, shims, housings, adhesives and brackets.

A useful historical example lies in aviation instrument practice itself. Long before autonomy became fashionable, aviation learned that instruments and references could not simply be trusted because they had once been correct. Attitude indicators, compasses, gyros and navigation aids required disciplined checking because flight conditions, vibration and system interactions could degrade their truthfulness without producing obvious breakage. The relevance to Arctic unmanned systems is direct. What older aviation culture treated as instrument discipline, modern autonomy must treat as calibration discipline across a larger and more tightly coupled sensor stack. The difference is scale, not principle.

A second example comes from polar and extra-terrestrial robotics, where remote systems operated by organisations such as NASA have depended on painstaking calibration maintenance across environments that punish thermal and mechanical complacency. Mars rovers and helicopters do not inhabit the Arctic, but they offer a clear engineering lesson: sensors remain trustworthy only when the programme assumes that alignment and bias are living concerns rather than once-and-done achievements. Publicly discussed mission engineering repeatedly shows attention to how temperature, motion, time and environmental exposure alter the meaning of sensor output. The lesson worth borrowing is not romance about space. It is patience about drift.

A fair limitation belongs here. Not every small calibration shift matters equally, and not every operation needs laboratory-grade precision to remain safe and useful. Many systems are designed with tolerances broad enough to absorb minor misalignments or biases, especially when multiple references remain healthy. Some payload tasks can tolerate small pointing errors. Some navigation regimes remain perfectly adequate with modest drift so long as external corrections are frequent and recovery geometries remain generous. It would be a mistake to turn every calibration deviation into a crisis.

But the limitation only strengthens the deeper argument. Because many systems can absorb small error for a time, organisations become tempted to treat calibration as something the vehicle can quietly "work around".

Sometimes it can. Sometimes that success merely delays discovery until a harsher moment, when other aids degrade and the accumulated misalignment becomes operationally significant. The right lesson is therefore not panic. It is discipline about hidden dependency. A system that tolerates drift gracefully in one phase may become brittle in another if no one has tracked how much truth was being borrowed from healthier references.

This is where maintenance becomes inseparable from calibration. Maintenance is not only the replacement of worn parts or the repair of obvious faults. In Arctic operations it is also the preservation of reference integrity. A technician cleaning an optical window, checking a mount, inspecting a connector, examining ice traces, or reseating a payload is not merely restoring appearance. That technician may be protecting the system's ability to keep its own geometry straight. Trust, here, is not a soft word. It appears in visible behaviour: whether operators accept automated cueing without a confirming pass, whether a supervisor permits degraded-mode operation based on inertial and heading references, whether an aircraft can be relaunched after minor icing exposure without a prolonged suspicion that its sensor alignment may no longer deserve confidence.

The relation between calibration and autonomy is therefore moral as well as technical. Autonomous systems act on the world according to the quality of their internal correspondence with it. If the correspondence is stale, shifted or partly fictitious, the machine may continue acting with perfect procedural confidence while its underlying reference frame has already begun to drift. That is a dangerous form of elegance. A system can appear stable, controlled and intelligent while moving inside a geometry that no longer matches reality closely enough. The cold does not need to break the machine. It only needs to move the meaning of its measurements.

This is why good programmes look for indirect signs. They do not wait for catastrophic divergence. They watch for small disagreements among sensors, recurring post-flight alignment tweaks, changes in how often operators question automated outputs, or subtle increases in correction activity during familiar manoeuvres. They ask whether a vehicle that once flew smoothly through degraded conditions has begun behaving differently after a season of vibration, icing and transport. They compare logs across airframes rather than flattering themselves that each slight anomaly is unique. Calibration failure often announces itself as pattern before it announces itself as incident.

The quiet motif here is the hinge. A hinge can still open and close after it has shifted. The door still works, until one day the latch no longer meets cleanly and everyone realises the whole geometry had been moving for some time. Calibration in harsh environments behaves much the same way. The outputs still arrive. The aircraft still flies. The cue still points somewhere plausible. Then a reference disappears, a weather window narrows, an automated recovery depends on one more layer of internal truth, and the misfit suddenly matters. The hinge was already telling the story.

This makes calibration discipline a question of institutional maturity. Programmes that treat calibration as paperwork will often discover drift only when operators have already developed informal distrust. Programmes that take it seriously build routines around temperature-aware checking, post-icing suspicion, mount integrity, returned-parts examination and the willingness to ground or re-verify a system whose outputs remain superficially convincing. They understand that calibration is not merely about starting from the right numbers. It is about knowing when those numbers have stopped describing the physical arrangement honestly enough for autonomy to deserve its authority.

The unresolved test comes when the aircraft is still flying well enough to tempt continuation, the displays still look tidy, and GPS or another strong external reference begins to weaken. At that moment, what exactly is carrying the machine's confidence: a genuinely preserved relationship among compass, IMU, payload and airframe, or a history of small thermal and mechanical shifts that have never yet been forced to answer for themselves? That question is the one the Arctic keeps asking, often just after the easiest source of reassurance has dropped away.

Data quality as an operational variable

On 1 June 2009, Air France Flight 447 disappeared over the Atlantic after a chain of failures in which sensor data, automation logic and human interpretation no longer held together. The event did not occur in the Arctic, and it would be crude to turn it into a general parable for every unmanned system. Yet one lesson from that night is exact and transferable: data does not become useful simply because it is present. It must also be trustworthy, interpretable and honest about its own limits. The number still looked clean.

That is the central difficulty with data in snow and ice. A modern autonomous platform can gather astonishing amounts of it: optical imagery, infrared scenes, radar returns, lidar clouds, inertial traces, GNSS updates, power states, environmental telemetry, health messages and mission logs. The temptation is to treat this abundance as security. More inputs, more storage, more fusion, more machine confidence, and therefore more operational truth. The Arctic repeatedly exposes the weakness in that sequence. Data can be plentiful and poor. It can arrive on time and still mislead. It can remain internally consistent while becoming less representative of the world outside the machine. Quantity is not innocence.

This is why data should be treated less like a passive record and more like fuel. Show the analogy in its strict operational sense and it becomes immediately useful. Fuel is measurable, consumed in use, degraded by conditions, vulnerable to contamination, and decisive only if its quality matches the demands placed upon it. Data behaves in much the same way in harsh environments. A system begins with some confidence in its sensing and state estimates. As conditions worsen, some of that confidence is spent. Low contrast, thermal flattening, clutter, icing, latency, calibration drift and communication dropouts all degrade the quality of the data available for action. Name the issue and the argument sharpens: data quality as a consumable. Explain it, and the implication is severe. A mission can run short of trustworthy data before it runs short of electricity, airframe integrity or flight time.

That way of thinking changes what autonomy ought to mean. People often ask whether a system is autonomous as though the answer depends on how many decisions it can make without direct human input. A more serious question is whether the system can recognise when the informational basis for those decisions is thinning out. That is a very different test. It asks not whether the machine can act, but whether it knows when it should act less boldly. Show the operational contrast and the point becomes clear. One platform continues presenting precise tracks, confident classifications and tidy route solutions even as weather, drift and clutter degrade its inputs. Another begins to narrow its claims, broaden its uncertainty bounds, reduce mission ambition or request confirmation. The second system may look less glamorous. It is often more truthful.

Truthfulness, here, is not a metaphysical virtue. It is a concrete operational behaviour. A system that reports confidence honestly does several things that can be seen and tested. It changes its conduct when data quality worsens. It reveals which inputs it is relying on more heavily. It distinguishes estimated knowledge from directly observed knowledge. It records when confidence changed and why. It stops presenting false precision as a substitute for missing certainty. That last point matters deeply in the Arctic because harsh conditions often remove obvious cues while leaving numerical outputs intact. A machine can still show coordinates, tracks and detections long after the evidential foundation beneath them has become less trustworthy. The numbers remain. The meaning thins.

A useful historical anchor lies in aviation itself. Modern flight has long depended not merely on instrument availability, but on disciplined interpretation of instrument quality, agreement and failure. Pilots are taught to distrust certain readings when sources disagree, when icing affects inputs, or when automation begins behaving in ways that imply uncertainty rather than clarity. The lesson is larger than cockpit procedure. It is that data is not a monolith. It has lineage, condition and quality. Arctic autonomous systems deserve to be judged by the same seriousness. If a drone's navigation or perception outputs degrade in meaning but not in visual neatness, the problem is not only technical. It is epistemic. The machine has failed to communicate what kind of knowledge it still possesses.

A second example comes from planetary robotics, particularly missions operated by NASA, where remote vehicles must act under severe uncertainty and where communication delays make it impossible to pretend that a human will always step in at the decisive moment. These systems have often been praised for their autonomy, but the more revealing quality is usually their caution about what the data can support. Hazard assessment, navigation and activity planning in such missions depend not merely on collected data, but on whether the system and its operators understand where that data is thin, delayed, noisy or context-bound. The relevance to Arctic operations is plain. Remote, harsh environments reward systems that know the difference between having information and having enough information to deserve confidence.

This distinction becomes especially important when multiple data streams begin agreeing for the wrong reasons. A visible camera sees weak contrast. Infrared sees weak separability. Lidar sees atmospheric clutter. GNSS remains numerically available but less helpful than expected. Inertial estimates stay smooth because they are leaning harder on prior motion. The fused output may look coherent. The display may remain calm. The operator may feel reassured because nothing on screen appears dramatically wrong. Yet the calm may be purchased by the system smoothing over uncertainty rather than confronting it. That is contaminated data in the operational sense: not fake, not absent, but degraded in a way that matters to judgement. Fuel with water in it may still fill the tank. Data with hidden weakness may still fill the display.

Contamination, therefore, is one of the most useful metaphors in this domain, provided it is handled precisely. Data becomes contaminated when it carries more ambiguity, error or unacknowledged bias than the system's presentation suggests. A lens fogged slightly, a radar return thickened by clutter, an IMU carrying a thermal bias, a gimbal shifted by icing, a battery state estimate detached from actual deliverability, a map no longer matching surface reality as closely as assumed - each may feed information into the system that is not wholly false and not clean enough to trust without qualification. The danger lies in the mixture. Pure failure is easier to manage than impure confidence.

This is why practical questions matter more than slogans about artificial intelligence or all-weather capability. When a system presents a target, a route, a landing cue or an energy estimate, what does it also report about the quality of the underlying data? Does it say which sensors are carrying the decision? Does it flag disagreement among them in a way that affects authority rather than merely decorating the interface? Does it distinguish a contact that is repeatedly confirmed from one inferred through fragile fusion? Does it expose whether confidence has fallen because the scene changed, the sensor degraded, or the system has shifted to a fallback estimation mode? These are not philosophical niceties. There are ways of asking whether the machine knows what it does not know.

One of the most revealing signs of epistemic discipline is whether uncertainty changes behaviour soon enough. A platform that displays a low-confidence warning but continues its nominal plan unchanged is not necessarily disciplined. It may simply be confessing weakness while refusing to honour it. A stronger system connects confidence to conduct. It climbs, slows, widens margins, drops non-essential tasks, abandons precision functions, or returns early. The reader need not think of these changes as evidence of timidity. They are better understood as operational honesty. A mission conducted on worsening data should not look identical to a mission conducted on strong data. If it does, the system is hiding something - perhaps from its operators, perhaps from itself.

That honesty also has an institutional dimension. Organisations often reward demonstrations of autonomy that look smooth, decisive and self-assured. Clean maps, stable tracks and confident overlays flatter procurement and policy audiences. A system that frequently displays uncertainty, narrows its claims or requests confirmation can appear less advanced to the untrained eye. Yet in Arctic operations, the opposite is often true. The more mature system is the one willing to expose the degradation of its own information state. It does not sell certainty cheaply. It preserves it for the moments when it has actually been earned. Marketing language tends to admire outputs. Operational seriousness should admire candour.

A fair limitation deserves acknowledgement. No system can report every aspect of data quality perfectly, and no operator can absorb unlimited layers of uncertainty notation in the middle of a demanding mission. Too much metadata can become its own confusion. Some missions also require action under conditions of partial knowledge, and a platform that constantly foregrounds ambiguity without prioritising it can become operationally paralysing. The criticism is real. An epistemic discipline badly designed can turn into self-conscious clutter.

But the limitation does not rescue overconfidence. It argues for selective, consequential transparency. The right issue is not whether every uncertainty can be shown. It is whether the uncertainties that matter operationally are shown in ways that alter behaviour. That means systems should expose the state of the data where it bears on route freedom, collision avoidance, target identification, landing authority, power reserve, navigation integrity and mission continuation. It means not every doubt needs a headline, but the important doubts must not be buried beneath polished graphics and optimistic defaults. The Arctic is not offended by incomplete knowledge. It is offended by knowledge pretending to completeness.

This has consequences for how data should be stored and reviewed after missions as well. If data quality is consumable, then post-flight analysis should not ask only what the aircraft saw or recorded. It should ask when the quality of that information changed, whether the system recognised the change, how operators responded, and whether the machine's self-description matched what later analysis reveals. A good log, in this sense, is not just a diary of events. It is a record of informational condition. Did confidence drop before the manoeuvre changed, or after? Did the machine narrow its authority when sensor disagreement rose? Did operators override because they distrusted the aircraft's confidence logic, or because the aircraft failed to surface the issue clearly? Data about data becomes indispensable once autonomy is judged not by marketing claims but by evidential conduct.

The fuel metaphor returns usefully here because it also clarifies planning. No responsible crew plans a mission on nominal fuel alone without regard to reserve, burn rate, quality and contingencies. Data deserves parallel treatment. A mission should not be built only around the best-case performance of its sensors and estimators. It should ask what happens when data quality degrades partway through, what reserves of certainty exist in other sensors or procedures, what forms of contamination are likely, and what operational rules apply when the informational reserve begins to shrink. The most honest systems are those that act as though data can be spent, spoiled and rationed.

The quiet motif in all this is the gauge. Operators love gauges because they promise that a hidden state has been made visible. Fuel gauges, battery indicators, confidence bars, health lights, track quality readouts - all try to condense complexity into something graspable at speed. The problem is not the gauge itself. The problem is when it flatters by compressing a contested reality into a calm-looking signal. A trustworthy system does not merely show a gauge. It ensures the gauge remains tied to the real condition of the thing being judged, and it changes the machine's conduct when the reading becomes less secure. A gauge without behavioural consequence is often just decoration.

The phrase "epistemic discipline" may sound grander than the machinery it describes. In practice it is made of very plain virtues. The system distinguishes between sensing and estimating. It records provenance. It reports disagreement. It avoids false precision. It degrades gracefully. It helps operators understand why confidence has changed. It refuses to let a tidy output stand in for untidy evidence. None of these qualities makes for flamboyant sales literature. All of them matter in snow and ice, where the scene itself is often quietly hostile to certainty.

That is why data quality should be treated as an operational variable rather than a technical afterthought. It changes across the mission. It interacts with weather, calibration, sensor condition, power state and communication limits. It can be guarded, spent, contaminated and, to a degree, replenished by better geometry, new references, slower tempo or more conservative rules. Once seen that way, the proper question shifts. The issue is no longer whether an autonomous platform carries enough data sources. It is whether the platform, and the organisation using it, can tell when the data has ceased to deserve the authority still being granted to it.

The unresolved test arrives at a familiar moment in harsh operations: the target still seems plausible, the route still looks open, the display still appears composed, and the environment has just become ambiguous enough that the machine's confidence should begin to contract. What matters then is not whether the platform can keep producing outputs. It almost certainly can. What matters is whether it admits that its informational fuel is no longer clean enough for the same level of ambition - and whether anyone reading the screen can tell the difference between a strong answer and a polished one.

Chapter 5

Navigation without GNSS Certainty

A vehicle can know its coordinates and still be lost. That is the Arctic navigation paradox in its cleanest form. A display may continue updating, tracks may remain smooth, route lines may still look sensible, and yet the deeper question - what exactly is anchoring those claims to the world outside - may already have become dangerously unsettled. The map remains. The reference thins.

That thinning happens faster in snow and ice than many designs admit. Visual landmarks disappear into white ground and pale sky. Surface features repeat until familiarity becomes unreliable. Altitude loses some of its intuitive meaning when the terrain beneath is flat, drifting, cluttered, or poorly distinguished from the atmosphere above it. An aircraft may keep flying with impressive composure while the basis for saying where it is, how high it sits, and what margin remains around it grows less certain. Navigation is not merely weakened by the Arctic. It is stripped of its easiest confirmations.

This is why navigation without assured satellite support cannot be treated as a narrow technical inconvenience. It is a systems problem from the beginning. Sensors matter, of course. So do inertial estimates, magnetic behaviour, map quality and software logic. But procedures matter just as much, because navigation in the High North is always partly a question of what the organisation is willing to assume, what it is willing to forbid, and how much uncertainty it is willing to price into a mission before launch. Once references become scarce, freedom itself becomes conditional.

The real tension lies there. Autonomy promises mobility, flexibility and delegated movement across wide, sparse territory. Harsh conditions push in the opposite direction. They reward structured routes, conservative envelopes, predeclared fallback behaviour and a stricter relationship between what the aircraft believes and what the mission will allow it to do on that belief. The more uncertain navigation becomes, the less honest it is to speak of autonomous freedom in expansive terms. Arctic operations do not abolish initiative; they charge for it. The bill is paid in wider buffers, earlier turn-backs, simpler paths and a lower tolerance for elegant guessing.

That matters because navigation error in the Arctic does not usually fail theatrically at first. It leaks into the operation by degrees. A route that looks neat on screen begins demanding more correction than it should. A loiter pattern feels slightly offset from expectation. A recovery approach depends more heavily on one remaining trustworthy source than anyone would prefer. An inertial solution that seemed calm on shorter flights begins to show the accumulated burden of endurance. A magnetic reference that was good enough when other cues were abundant becomes awkward when those cues degrade. Trust, here, becomes visible in behaviour: whether operators widen margins without being told, whether supervisors refuse shortcuts near terrain or ice edges, whether planners reduce ambition because they no longer believe the machine's certainty is worth the route drawn around it.

The stakes therefore extend beyond position error in the abstract. Navigation quality shapes how a programme defines safe flight itself. If reference is weak, then altitude policy changes, corridor design changes, divert planning changes, recovery expectations change, and the meaning of acceptable autonomy changes with them. A platform that cannot prove where it is with enough integrity may still remain airborne and nominally functional, yet operationally it has already become a more fragile thing. The Arctic turns location into governance.

The path through this chapter follows that governance problem as it emerges from the environment rather than from the software manual. It begins where all northern navigation difficulties begin, in whiteout and reference scarcity, where the world stops offering reliable edges and external truth becomes harder to borrow. It then moves across the different ways systems try to hold their place - inertial, magnetic and map-based - before tightening into route structure, geofences and the deliberate narrowing of freedom that uncertainty demands. From there the argument settles on integrity itself: not whether the machine can continue producing a position, but whether it knows when that position has ceased to deserve full authority.

A useful way to hold these questions together is to think of a handrail in darkness. A handrail does not tell you everything about the building, but it offers a reference that lets movement continue without pretending the whole space is plainly visible. Arctic navigation works best when it is designed with that same modesty. A system should not need omniscience to remain useful. It should need enough verified reference, enough honest fallback, and enough discipline about what must narrow when the easy picture of the world has gone.

The harder question, then, is not whether a vehicle can navigate after certainty begins to erode. Many can, for a time. The real question is what exactly the system, its operators and its rules are prepared to stop doing once location becomes less a fact than a wager - and whether they will recognise that moment before the map on the screen begins to look more truthful than the world beneath the aircraft.

Whiteout navigation and reference scarcity

On 28 November 1979, Air New Zealand Flight 901 flew into the slopes of Mount Erebus on Ross Island, Antarctica. The accident had several causes, and it would be careless to force it into a single moral. Yet one lesson from that disaster belongs squarely to Arctic navigation: a landscape can remain present while ceasing to provide reliable reference. The aircraft was not moving through emptiness. It was moving through a visual world that no longer yielded the cues required for safe judgement. The horizon had gone.

That phrase matters because whiteout is often described too loosely, as though it were merely bad visibility by another name. Whiteout is harsher than that. It is a condition in which the ordinary scaffolding of navigation - horizon, shadow, edge, scale, slope, separation between surface and sky - begins to dissolve. Objects may still be seen. Distance may still appear estimable. A pilot or remote operator may still feel that enough information exists to continue. Yet the very references that make position, attitude and motion intelligible have thinned or flattened into something less trustworthy. A machine operating in such a scene may continue to report location, altitude and track with outward composure. The question is what those claims are still anchored to.

That is the beginning of reference scarcity. It does not require full blindness, nor the dramatic blankness people imagine when they hear the word whiteout. Near-whiteout can be more treacherous because it preserves enough scene structure to encourage belief. Snowfields still look like surfaces. Ice ridges still look like edges. A distant shoreline or hummock may seem available as a check on position or heading. The trouble is that these fragments of reference can repeat, blur or move in meaning as light and weather shift. A drone crossing a pale, patterned landscape may appear to be navigating over stable features when in fact the features offer far less uniqueness than the system assumes. The map says one thing. The world offers echoes.

This is why Arctic navigation becomes difficult even before one begins talking about outright loss of GNSS or deliberate interference. The problem starts at the level of the scene itself. Show the aircraft low over sea ice, snowfield or frozen littoral under flat light, and the navigation burden changes immediately. Visible cues weaken. Terrain features repeat. Surface relief becomes harder to read. Altitude above ground becomes less intuitive because the eye - or the vision system - loses the textural gradients that usually support depth and motion judgement. Name the issue precisely and it becomes more than weather: reference poverty. Explain it and the operational implication is stark. A vehicle may still know roughly where it is while becoming less certain about how it sits in relation to what matters nearby.

This is where mapping quality enters the argument, not as a comforting substitute for the outside world, but as another dependency exposed by it. Maps are often imagined as fixed truth beneath moving uncertainty. In the Arctic, that confidence needs trimming. Some terrain is dynamic. Snow cover alters surface appearance and sometimes effective geometry. Sea ice is mobile by definition. Shorelines, ridges, drifts, melt features and seasonal changes can all weaken the correspondence between mapped representation and present operational reality. Even where maps remain broadly sound, their resolution and update quality may not support the level of proximity or low-level manoeuvre that a mission silently expects. Show a route line laid cleanly across a snow-covered expanse and the question becomes unavoidable: how much of this neatness belongs to the map, and how much belongs to the actual surface the aircraft will meet today?

A useful historical example lies in polar aviation more broadly, where pilots and expeditions have long known that navigation over snow and ice is not just a matter of carrying instruments into a blank place. It is a matter of recognising how little the environment offers back by way of confirmation. Antarctic and Arctic flight practice developed its own caution precisely because visual impressions over white terrain can be deeply misleading even when some visibility remains. The significance of the Erebus accident rests partly here. It showed with terrible force that a scene can seem interpretable while concealing fatal loss of reference. For unmanned systems, the lesson is neither nostalgic nor symbolic. It is direct: the sensor suite and software inherit the same poverty of cues, only translated into pixels, returns and state estimates.

A second example comes from modern drone operations over sea ice and snow-covered terrain described in scientific and inspection contexts, where crews often discover that navigation and imaging degrade together. A platform may hold a route accurately in coordinate terms while operators lose confidence in what the aircraft is actually above, how high it truly is above the relevant surface, or how much clearance remains around an intended path or landing area. The important point in these reports is not that the systems become useless. It is that certainty narrows unevenly. Geographic position may remain acceptable for one purpose and inadequate for another. A route that is perfectly safe at one altitude becomes questionable at another. A landing zone that appears mapped and obvious from planning tools becomes visually unconvincing in the actual light. Navigation ceases to be one thing.

Altitude estimation is where this layered uncertainty becomes especially dangerous. People often speak casually of altitude as though one number solved the matter. In Arctic operations, altitude must be qualified. Altitude above mean sea level is not altitude above terrain. Altitude above a mapped surface is not necessarily altitude above the true operational surface, especially where ice, snow, drifts or dynamic maritime conditions complicate the picture. A radar altimeter, lidar, barometric estimate or fused state may each tell a slightly different story depending on environment, attitude, clutter and calibration. Show a low-flying aircraft over a smooth snowfield or broken sea ice and the point sharpens. A few metres of unrecognised error may not matter high in transit and may matter enormously in low-level survey, approach or obstacle avoidance. Reference scarcity turns vertical navigation into a live systems question rather than a single-instrument readout.

This is one reason Arctic navigation cannot be reduced to sensors alone. Sensors matter, but their meaning is carried by software and bounded by procedure. A camera may lose contrast. A range sensor may encounter ambiguous returns. An inertial estimate may drift slowly. A map may be good in a strategic sense and poor in a tactical one. Software must decide how those facts are weighed. Procedures must decide what freedoms remain when the weighting grows doubtful. That is why the navigation problem in whiteout is not solved simply by adding another instrument. A new sensor may contribute useful evidence, but unless the system knows how reference scarcity changes the authority of each input - and unless operators know what that shift requires of them - the new sensor may merely add another channel through which uncertainty travels.

The broader stakes become visible in behaviour. Trust in navigation is not a sentiment floating above the operation. It appears when operators let the aircraft continue low over feature-poor terrain without demanding a wider buffer. It appears when planners accept a narrow route corridor because they believe map, altitude and position solutions still deserve that precision. It appears when a recovery proceeds without extra orbiting or widened margins because the system's reference state seems intact. When that trust weakens, the signs are equally concrete: routes rise higher, turns widen, descent points move farther out, recovery options multiply, operators ask for confirming passes, and sorties that looked feasible on paper quietly shrink in practice. Efficiency in this environment is often little more than the rate at which good reference can be converted into safe movement.

A fair limitation deserves acknowledgement. Not every whiteout or near-whiteout condition makes navigation impossible, and not every Arctic mission depends equally on rich visual reference. A platform operating at safer altitude over known corridors, with conservative route design and strong external position support, may remain entirely viable even when the scene below is visually poor. Some sensors and procedures can preserve enough integrity for useful operations in degraded conditions. It would be careless to imply that every loss of contrast demands immediate retreat.

But that limitation only strengthens the central argument. Where navigation remains workable under whiteout-like conditions, it is usually because the system and its rules have already narrowed what the mission asks of it. The aircraft flies higher, straighter, slower, farther from obstacles, or under stricter corridor logic. The organisation has converted uncertainty into procedure before the sortie begins. That is the reconciliation worth keeping. Whiteout navigation is not defeated only by better hardware, nor mastered only by better piloting or software. It is managed by an honest relationship between sensing, mapping, altitude strategy and rules of use.

This honesty becomes especially important where terrain features repeat. Repetition is one of the Arctic's quietest threats. A machine seeking correspondence between scene and map, or between successive observations, does badly when the landscape offers long stretches of near equivalence: parallel ridges, uniform snowfields, repeated pressure features, indistinct coastlines, and shadows too weak or too unstable to help. Human beings are vulnerable to this too, though they may not admit it. One pale hollow resembles another. One ice edge seems much like the next. A navigation stack built on matching and cross-checking can become overconfident when the environment provides multiple answers that all appear plausible. Scarcity is not only having too little reference. It is having reference that is too repetitive to constrain error strongly.

The quiet motif here is the horizon. It is not merely a line in the distance. It is a permission structure for orientation, motion and scale. When the horizon is clear, many small uncertainties remain tolerable because the world keeps offering free correction. When it disappears, or becomes ambiguous, the rest of the navigation system inherits a burden it may or may not have been designed to carry. Cameras lose an anchor. Operators lose a check. Altitude feels less intuitive. Surface motion becomes harder to infer. Mapping becomes more heavily relied upon just as the scene becomes less willing to confirm it. The horizon is not everything. Its absence changes everything.

This is why safe flight envelopes in the Arctic must be treated as navigation decisions as much as aerodynamic ones. A system may be capable of operating in a certain wind band, at a certain altitude, or along a certain route in physical terms. The harder question is whether it can do so while reference remains strong enough to justify that freedom. A route that is technically flyable may be navigationally dishonest if whiteout and repeated terrain strip away enough confirming cues. A low-altitude survey that is aerodynamically routine may become unacceptable because altitude above true surface can no longer be known well enough. The envelope shrinks not because the airframe weakens, but because the truth available to it does.

The unresolved test, then, is not whether the machine can keep producing position, altitude and track while the world turns pale and repetitive. Many machines can. The sharper question is whether those outputs still deserve the same authority once the horizon has dissolved, the map has become more aspirational than immediate, and the scene below offers too few unique references to correct a growing error before it enters the flight path itself.

Inertial, magnetic, and map-based methods

On 3 August 1958, USS *Nautilus* passed beneath the North Pole, doing something that had once seemed almost absurd: navigating with confidence in a place where the familiar certainties of compass-bearing and ordinary chart logic grow strained. The submarine's achievement is often remembered as a geopolitical and technological triumph. It should also be remembered as a navigation lesson. Near the pole, the world does not yield reference in the way lower latitudes teach people to expect. Instruments still work, but not in the same moral climate. North becomes difficult.

That phrase deserves more weight than it usually receives. Navigation at high latitude is not simply ordinary navigation conducted in cold weather. It is navigation under altered reference conditions, where each method carries a different kind of weakness and none should be mistaken for an all-weather, all-duration truth machine. Inertial systems drift. Magnetic systems become awkward or unstable in ways that are geographically and locally specific. Map-based methods depend on the world remaining distinctive enough, stable enough and well represented enough to match. The aircraft may still know something useful. The hard question is what kind of knowledge it still possesses, and how quickly that knowledge decays.

This is why redundancy in Arctic navigation cannot be understood as mere quantity. Adding several methods does not automatically create safety. The value lies in difference: in understanding how each method fails, how quickly it fails, and what sort of lies it is likely to tell when conditions turn against it. Show the operational problem plainly and the point sharpens. A vehicle loses confidence in one source of position and leans more heavily on another. That is only a virtue if the second source is not vulnerable to the same environment, the same assumptions, or the same kind of hidden error. Name the principle and it becomes more than engineering folklore: failure diversity. Explain it, and the implication becomes severe. Real redundancy comes from disagreement that teaches, not agreement that flatters.

Inertial navigation appears, at first, to be the most dignified answer to Arctic uncertainty. It does not need stars, landmarks or external transmissions in the ordinary way. It measures motion and integrates it into a story about where the vehicle is going and where it has gone. This independence gives inertial systems their authority. It also gives them their burden. Show an aircraft flying cleanly under good initial alignment and the inertial solution can look almost sovereign: smooth, stable, unflustered by whiteout, darkness or reference-poor surfaces. Name the underlying weakness and the picture changes at once: accumulated error. Explain it and one sees why inertial systems are admired and feared at the same time. Small biases in acceleration or rotation do not remain small when repeatedly integrated across time. The error may grow quietly, almost elegantly, until endurance, distance or loss of external correction turns a tolerable drift into an operational fact.

That is why an INS can seem stable until the mission becomes long enough to ask what stability has been bought with. This is one of the Arctic's characteristic cruelties. Short sorties flatter inertial solutions because little time exists for drift to become legible. Longer endurance begins to expose the difference between smoothness and truth. A platform can hold a tidy track history and still be moving away from its real position by increments that matter only later, often when recovery options are fewer and the system has already become more dependent on its own estimate. Trust, here, becomes visible in behaviour. Operators stop trimming routes tightly around hazards. Supervisors insist on larger return margins. Mission planners shorten legs or add confirmation points not because the aircraft has failed, but because its inertial confidence has grown too serene to be taken literally.

Magnetic methods carry a different burden. Their attraction is obvious. They are light, familiar and in many systems deeply integrated into heading estimation. But high latitudes expose how conditional magnetic usefulness really is. The Earth's magnetic field becomes more difficult to live with as inclination steepens and local disturbances matter more. Add the realities of aircraft installation - current-carrying wires, motors, structural materials, payloads, heaters, connectors, and changing power states - and the magnetic picture can become even less cooperative. A compass is not simply reading the planet. It is reading the planet through the airframe, its electrical life and its environment. In the Arctic, that mediation grows less forgiving.

This does not mean magnetics become useless. It means they must be read with unusual caution. Show a small unmanned aircraft operating in severe cold with changing payload states and one can imagine the pattern well enough. A heading solution that seems acceptable during benign flight becomes more suspect when GNSS quality worsens, visual cues fade, or a long endurance segment increases dependence on internal reference. A small magnetic bias, invisible while other sources are strong, begins to steer decisions once the system leans on it more heavily. Name the issue and the engineering stakes become practical: conditional authority. Explain it, and the implication follows. A magnetic input should rarely be treated as a free-standing truth in high latitudes. It is better understood as one witness whose credibility depends on context.

The history of polar navigation offers a persistent reminder of this. Explorers, aviators and submariners operating near the poles learned early that magnetic compasses could become unreliable or awkward enough that alternative methods, including gyro and inertial techniques, took on unusual importance. The relevance of *Nautilus* rests partly here. Its under-ice navigation was not simply a victory of machinery over geography. It was proof that high-latitude movement forces navigators to choose reference systems according to how each one fails. That remains true for autonomous systems, even if their interfaces are cleaner and their mathematics more elaborate.

Map-based methods appear to promise a different kind of confidence: if the environment can be matched to a stored representation, the machine may recover its place without depending solely on satellite signals or inertial continuity. Sometimes that works impressively. When terrain is distinctive, when maps are good, when the sensing geometry is kind, and when the environment has not changed too much, map matching can provide the kind of external correction inertial systems crave. Yet the Arctic is unusually good at exposing the limits of this promise. Snow cover alters visual and geometric appearance. Sea ice moves. Coastal zones can present repeated forms at the scale that matters to autonomous vehicles. Low contrast weakens optical matching. Surface clutter or atmospheric interference can complicate active-sensor correspondence. The map remains. The match may not.

This is why map quality matters in two senses at once. It matters whether the underlying representation is accurate, current and resolved finely enough for the mission. It also matters whether the environment still offers the sensor a scene that can be matched to that representation with honest confidence. Show a drone trying to infer its position over a broad, pale, repeating surface and the problem becomes immediate. Even if the map is fundamentally sound, the live scene may not contain enough distinctive features to constrain the solution strongly. Name that weakness and it becomes more exact than simple "poor mapping": low correspondence strength. Explain it, and one sees why map-based navigation can be both useful and fragile. It works best not when a map merely exists, but when the world is willing to resemble it in operationally helpful ways.

A second named example helps frame this more broadly. The 1979 Mount Erebus disaster, though not a simple navigation-method story, remains instructive because it showed how dangerous it is to move through a polar environment with a false sense of reference. Visual interpretation, route logic and environmental deception combined in a setting where the world no longer provided the confirmations that ordinary flight habits encouraged crews to expect. The lesson for unmanned Arctic systems is not that they will repeat that event. It is that navigation cannot be judged solely by what the internal solution appears to say. One must also ask what external world is still available to constrain and correct it.

The attraction of combining inertial, magnetic and map-based methods is therefore entirely reasonable. Each can cover part of what another lacks. Inertial systems provide continuity when external reference fades. Magnetic inputs can support heading estimation when used carefully and contextually. Map-based methods can re-anchor the solution when the scene and representation truly correspond. The problem comes when designers or operators forget that these methods may weaken in patterns that overlap operationally even when they differ mechanically. A poor visual scene can weaken map matching. Long endurance can expose inertial drift. High-latitude and installation effects can thin magnetic reliability. The system still has three methods. Its real navigation freedom may already be shrinking.

This is where redundancy design becomes intellectually serious. The question is not simply whether the platform possesses several navigation aids. The question is whether the aids fail with enough difference that one becomes more trustworthy when another begins to decay. A smooth inertial track is helpful if a credible external correction remains available. A magnetic heading aid is helpful if its distortions are known, bounded and not mistaken for certainty. A map-based fix is helpful if the scene really constrains it rather than merely allowing a plausible-looking correspondence. Redundancy that does not account for these conditions is not false exactly. It is over-advertised.

A fair limitation deserves emphasis. Some platforms and missions can make excellent use of these methods together, especially when routes are structured, endurance is moderate, environmental conditions are understood, and external aids are not entirely absent. High-quality inertial systems can hold useful truth for long periods. Careful magnetic integration can still contribute. Good map products combined with appropriate sensing can materially strengthen navigation. It would be lazy to imply that high-latitude operations are condemned to chronic uncertainty simply because GNSS confidence is reduced or reference is sparse.

But the counterpoint only clarifies the main argument. Success comes not from any one method becoming heroic, but from the system knowing what each method can still justify under present conditions. That is the discipline worth valuing. A strong navigation stack is not the one with the longest list of methods. It is the one whose methods disagree informatively, correct one another honestly and surrender authority visibly when their assumptions no longer hold. Arctic operations reward this modesty because the environment is unusually hostile to silent overconfidence.

The quiet motif here is the needle. A compass needle suggests a decision by direction, a simple answer to a simple question. Yet in high latitudes the needle becomes an emblem of a larger truth: every navigation method points, but not every one points with the same kind of honesty, and not all errors announce themselves with equal drama. Inertial drift does not twitch theatrically; it accumulates. Magnetic weakness does not always look like nonsense; it may look almost right. Map matching does not always fail noisily; it may settle on a plausible but weakly constrained answer. The art lies in knowing which kind of almost-rightness the system is living with.

That is why autonomy without GNSS certainty cannot be built on a method alone. It has to be built on interpretation, procedure and control of ambition. What routes are acceptable if inertial error may accumulate? What headings deserve trust if magnetic conditions are thin? What corrections can a map offer if the scene has become repetitive or altered? What margins must expand as each method's weakness becomes more relevant? These are not auxiliary questions. They define whether the aircraft's navigation solution is a usable basis for action or merely a tidy story told under pressure.

The unresolved test comes late enough to matter and early enough to trap the unwary: the aircraft is still flying well, the navigation displays remain smooth, and the mission appears entirely manageable. Only the duration has grown, the visual scene has become less distinctive, and one more source of external correction has weakened. At that moment, which method is the system really following - and does anyone watching understand how much of that confidence is being carried by a needle, a drift, or a map that may already be less specific than the aircraft's behaviour implies?

Geofencing, route structure, and safe corridors

On 13 January 1982, Air Florida Flight 90 departed Washington National Airport and, within moments, struck the 14th Street Bridge before crashing into the Potomac. The causes lay in icing, performance and judgement rather than in Arctic drone routing. Yet the deeper lesson travels north with surprising force. Freedom of movement is never a pure technical good. It is only as safe as the margins around it, the conditions it assumes, and the discipline with which those conditions are enforced. A route is not a line. It is a promise.

That promise becomes unusually expensive in the Arctic. People often imagine autonomous systems as embodiments of freedom: uncrewed vehicles slipping across distance, adjusting course intelligently, going where people would rather not go, and doing so with the fluidity of software rather than the stiffness of doctrine. Harsh polar conditions push back against that image. When visibility collapses, maps become less exact than they look, reference points repeat, and recovery options narrow, the value of autonomy no longer lies in open-ended manoeuvre. It lies in disciplined movement within boundaries. The box matters.

That is where geofencing enters the story, though the term can sound more trivial than the work it performs. A geofence is often described as a digital boundary, a region within which an aircraft may operate and beyond which it must not pass. Technically that is true. Operationally it is more than that. A geofence is a translation of uncertainty into rule. Show the practical case and the meaning sharpens at once. A mission planner knows that terrain reference is poor along one shoreline, that communications degrade behind a ridge, that recovery options vanish beyond an ice edge, or that a certain approach corridor becomes difficult to verify in flat light. Rather than trust the aircraft or the operator to improvise perfectly in the moment, the planner builds constraint into the mission space itself. Name the principle and the engineering becomes governance: uncertainty is being converted into permitted and forbidden movement. The implication is direct. Safe autonomy depends not only on what a platform can do, but on what it is never allowed to attempt.

This is why structured routes deserve to be understood as part of navigation rather than as a timid afterthought imposed on it. In a benign environment, route freedom can be generous because the penalties for small error are low and the opportunities for correction abundant. In the Arctic, route freedom is often the first luxury to disappear. A corridor may be preferred not because it is the shortest path, but because it preserves altitude options, keeps the aircraft over more trustworthy surface references, avoids the most deceptive backgrounds, maintains communications geometry, or stays within the radius of plausible recovery. Show a neat route laid across the polar map and the temptation is to see efficiency. Look harder and one often sees something sterner: a path chosen because the surrounding space is not equally survivable.

This distinction matters because many people hear the word geofence and imagine legal airspace or bureaucratic exclusion. Those uses are real, but the Arctic's more interesting fences are operational. They may not exist because a border or regulation demands them. They exist because the aircraft's confidence, if allowed to expand too far beyond them, would outrun the quality of reference, support or recoverability available on the ground and in the air. A safe corridor is therefore not only a navigation aid. It is a risk argument expressed spatially. It says: here the system still knows enough, can still return plausibly, can still be supervised honestly. Beyond this line the mission may still be imaginable, but it ceases to be responsibly governable.

A useful historical example lies in instrument flight and controlled airspace more broadly. Aviation learned long ago that freedom in degraded conditions is rarely increased by pretending every patch of sky carries equal safety value. Published procedures, approach corridors, holding patterns and altitude constraints are all, in a sense, structured responses to uncertainty. The Arctic radicalises this logic. Instead of a well-served network of alternates, radars and published infrastructure, one may have sparse support, deceptive surfaces, intermittent communications and few places where a vehicle can safely divert or be recovered. The lesson carried over from conventional aviation is not obedience for its own sake. It is that disciplined pathing is often what turns uncertain movement into acceptable movement.

A second example comes from publicly described polar and remote drone operations, where flight plans are often shaped less by the nominal reach of the aircraft than by what lies beneath and beyond the route. Scientific and inspection teams operating over sea ice, snow-covered terrain or cold littoral waters repeatedly discover that certain legs are operationally different from others even when the map distance is similar. A path over more legible terrain, better recovery access or stronger communications may be longer on paper and safer in truth. The significance of these cases lies not in dramatic incidents but in their quiet regularity. Arctic route design is often an exercise in refusing the geometrically neat answer in favour of the one with a more defensible margin.

This is where the phrase safe corridor earns its severity. A corridor is not merely a lane. It is a bundle of assumptions about reference quality, navigation confidence, energy reserve, wind tolerance, link geometry, obstacle margin and recovery plausibility. The corridor can therefore widen or narrow as conditions change. Flat light, blowing snow or degraded navigation integrity may force a broader separation from hazards or a higher altitude band. Poor communications may contract the usable lateral space. A suspected sensor issue may shorten the route altogether. Name the principle and its meaning becomes plain: route structure is conditional. Explain it and the operational implication follows. A corridor that looks generous in calm, clear planning may be functionally tight once the Arctic starts charging for uncertainty.

That conditionality is exactly why "freedom" for autonomy so often shrinks in harsh environments. This shrinking is not evidence that the platform has failed, nor that autonomy was oversold in some simple sense. It is evidence that the mission is being brought into honest proportion with the information and recoverability actually available. A vehicle may be physically capable of ranging far beyond its core corridor. It may have the endurance, the nominal control authority and the software sophistication to do so. But if its navigation confidence would degrade beyond a certain point, if divert sites become implausible, if communications thin, or if maintenance recovery becomes punishing, the area within which autonomy can be responsibly delegated contracts. Freedom, in such cases, is not removed by timidity. It is priced by reality.

This pricing shows itself in visible behaviours. Operators start rejecting seemingly modest route deviations because those deviations cross out of well-understood geometry and into a zone where recovery would be ugly. Supervisors insist on corridor discipline even when the payload task lies just beyond it. Mission planners add doglegs that keep the aircraft inside link coverage or within a safe drift-back option rather than draw the shortest line. Recovery points are chosen less for convenience than for the way they anchor the route's final uncertainties. These are not abstract displays of caution. They are the operational form of trust: evidence that a programme understands where its aircraft still knows enough to be allowed a certain degree of agency.

Geofences become especially valuable when navigation quality is not simply good or bad, but variable. Arctic operations are full of such variability. A route segment may begin over trustworthy ground, cross a more ambiguous surface, then end near an area where map correlation is poorer and weather effects on perception are stronger. Rather than treat the whole mission space as uniform, a mature programme assigns different rules to different regions. One zone may permit standard route following, another may require higher altitude, another may forbid autonomous descent, another may trigger mandatory return if navigation integrity worsens. The border on the map is therefore standing in for a border in confidence. Space becomes epistemic.

This has an organisational consequence that is often overlooked. Geofencing and structured routes encode institutional judgement before the sortie begins. They decide, in advance, how much uncertainty the organisation is prepared to tolerate over different terrain and in different states of system health. That makes them governance tools as much as navigation tools. A fence is a decision taken early enough to resist the seduction of the moment. An operator watching a target drift temptingly just outside the permitted corridor may be capable of making a clever exception. The point of the corridor is that the organisation has already judged the exception too expensive to normalise. This matters in the Arctic because difficult conditions reward pre-commitment over improvisational bravado.

A fair limitation belongs here. There are circumstances in which rigid geofences and narrow route structures can become liabilities. If a fence is based on stale assumptions, poor mapping or an overly cautious doctrine, it may waste mission opportunity and create its own operational awkwardness. Emergencies and unusual conditions sometimes justify temporary freedom that routine rules would deny. Some platforms, especially those supported by stronger sensing, stronger communications and more generous recovery options, can safely tolerate broader manoeuvre than more austere systems can. It would be foolish to praise restriction as a virtue in itself.

But the limitation does not weaken the main argument. It clarifies the standard. The question is not whether freedom is good and fences bad, or vice versa. The question is whether the spatial rules of the mission match the quality of navigation, perception, support and recovery actually available. A badly chosen corridor can indeed be wasteful. A well-chosen one converts uncertainty into manageable geometry. The reconciliation lies in recognising that routes and fences should move with evidence, not with fantasy. They must be strict enough to matter and flexible enough to reflect real conditions rather than inherited habit.

One of the most useful insights here is that route structure often preserves autonomy rather than suffocates it. A platform allowed to roam too widely in deceptive conditions may soon lose the confidence required for any meaningful self-direction at all. A platform held inside a corridor designed around its true strengths may operate with greater assurance, because the surrounding constraints reduce the chance that one bad estimate, one poor recovery geometry or one marginal communications loss will force an abrupt human takeover or early termination. Constraint, in this sense, is enabling. It gives the machine a smaller world in which its judgement remains more defensible.

The quiet motif in all this is the fence post. A fence post is unremarkable until visibility fades and distance grows, then suddenly the line it anchors becomes the difference between wandering and returning. So it is with geofences and corridors in Arctic autonomy. They are easy to mock when the weather is good and the system healthy. They become invaluable when reference thins, surfaces mislead, recovery choices contract and the temptation to believe the map over the world begins to rise. The post matters most when the line stops feeling decorative and starts feeling like memory laid down in advance.

This perspective also changes how one reads claims about autonomous adaptability. A system that advertises freedom to reroute, retask and optimise on the fly may sound more advanced than one bound to structured paths and fenced options. Sometimes it is. In the Arctic, however, the more advanced system may be the one that knows exactly when *not* to exercise that freedom. Adaptability is only useful if it remains inside a domain where the aircraft's knowledge and the organisation's recovery plan still overlap. The moment the aircraft can change its path more easily than the programme can safely justify the consequences, adaptability turns into a polished form of overreach.

The hardest question therefore arrives before any technical failure does. It appears when a mission looks almost easy: the aircraft is stable, the route is progressing, the target lies just beyond the planned corridor, and the displayed navigation solution still appears persuasive. What matters then is not whether the machine can physically leave the fence. Many can. The question is whether the corridor was drawn honestly enough that crossing it would mean more than a line violation - whether it would mean stepping out of the small, carefully built space where navigation uncertainty, recovery options and delegated control still belong to one another.

Integrity monitoring and decision thresholds

On 1 June 2009, Air France Flight 447 disappeared over the Atlantic after unreliable airspeed data and the system responses around it helped turn uncertainty into catastrophe. The accident was not an Arctic navigation case, and it would be simplistic to borrow it wholesale. Yet one lesson is painfully transferable: the most dangerous moment often arrives not when information disappears completely, but when a system continues acting as though its information still deserves full authority. The numbers stayed alive.

That is the core problem integrity monitoring is meant to solve. A navigation system should not merely produce position, velocity and attitude estimates. It should also ask, continuously and unsentimentally, whether those estimates remain trustworthy enough for the task being attempted. In severe environments, that second question is often the more important one. A vehicle over snow and ice can keep drawing a neat track across a display long after the evidential ground beneath that track has begun to weaken. GNSS quality may thin. Inertial drift may accumulate. Magnetic heading may become less reliable. Map correspondence may grow ambiguous. The solution can remain smooth while its right to govern the mission quietly erodes. The alarm should sound before the drift becomes visible.

Integrity monitoring is therefore not the same thing as navigation. Navigation tells the machine where it thinks it is. Integrity monitoring tells the machine how much that belief should be allowed to matter. Show the operational distinction, and the stakes become plain. A drone is flying a route over reference-poor terrain. Its position solution still updates normally. Yet one or more underlying checks begin to indicate that the solution is becoming less secure than the route structure, altitude, obstacle margin, or recovery plan can honestly tolerate. Name the principle, and the engineering becomes clear: protect the mission from overconfident location estimates. Explain it, and the implication follows immediately. A system with weak integrity logic may continue elegantly into danger. A system with strong integrity logic may become inconvenient earlier and survive to be useful again.

This is why integrity is best understood as a question of authority rather than mere accuracy. Accuracy matters, of course, but it is not enough. A navigation estimate can be somewhat inaccurate and still safe for a broad corridor at generous altitude. The same estimate may be wholly inadequate for a narrow approach, low-level inspection leg or constrained recovery. The issue is not simply whether the solution is wrong. It is whether the uncertainty around it has crossed a threshold at which the current operation no longer deserves permission. That threshold is the real hinge of autonomous behaviour in harsh environments. It decides when the system may continue, when it must simplify, and when it should stop pretending the difference does not matter.

This is also where the discussion leaves pure engineering and enters governance. Decision thresholds are sometimes presented as technical artefacts, as though sensor performance and algorithm design alone could discover them. They cannot. A threshold always encodes a view about acceptable risk. Show a vehicle allowed to continue low over a feature-poor, obstacle-rich corridor until navigation confidence drops to a certain point, and one is looking not merely at mathematics but at organisational judgement. Someone has decided how much uncertainty may be tolerated in exchange for mission completion. Someone has decided what counts as enough integrity for return, loiter, diversion, landing or abort. Name the issue properly, and it becomes unavoidable: thresholds are policies written into code.

That is why two systems with similar sensors can behave very differently and both still be acting "as designed". One programme may choose earlier withdrawal from precision manoeuvre because it values recoverability over collection completeness. Another may accept greater uncertainty in exchange for a broader operating window, perhaps because the consequences of abort are strategically, financially or logistically costly. Neither choice is purely technical. Each reflects a different institutional answer to the question: how much doubt is allowed before delegated action must shrink? The Arctic is merciless in exposing vague answers because it removes the spare margins that often conceal them elsewhere.

A useful historical example lies in aviation's long development of integrity concepts for instrument and satellite-supported flight. Modern aviation did not become safer merely by producing more precise navigation. It became safer by formalising when navigation precision and integrity were sufficient for specific phases of flight, and when they were not. The significance of this history is not bureaucratic. It is conceptual. The same position solution may be acceptable in one phase and unacceptable in another because the consequences of being wrong differ. Arctic autonomous systems face the same logic, only under harsher environmental uncertainty and often with thinner recovery options. Precision without integrity criteria is only numerical optimism.

A second example comes from remote and extra-terrestrial robotics, including missions managed by NASA, where systems must act in environments that cannot be corrected quickly by human intervention. Such systems succeed not only because they can estimate their position or state, but because they are built to narrow their own authority when the evidential basis for that estimate weakens. The lesson worth borrowing is not any specific algorithm. It is the disciplinary stance. A serious remote system does not ask only "Where am I?" It also asks "How dangerous is it to continue behaving as though I know?" That question is the beginning of integrity.

The practical mechanics of integrity monitoring vary, but the underlying logic is consistent. Good systems compare sources, watch for inconsistency, track the health of sensors, observe how uncertainty is evolving, and check whether current confidence still matches current mission geometry. An inertial solution that remains smooth may still be demoted if external corrections have been absent for too long. A GNSS-based solution may still be treated cautiously if geometry or signal behaviour suggests that nominal availability is masking weaker trustworthiness. A map-matching result may be accepted provisionally or discounted depending on scene distinctiveness. The important point is not that the machine computes more. It is that computation alters authority in time to matter.

This is especially important in Arctic conditions because degradation is often gradual rather than dramatic. If navigation simply vanished, most systems and crews would know what sort of emergency they were facing. The more difficult cases are the ones in which navigation remains present but less believable. A long-endurance aircraft may drift slowly enough that the solution looks stable. A magnetic heading input may remain near-plausible rather than clearly absurd. A map-based correction may fit weakly enough to reassure but not strongly enough to deserve the narrow corridor still being flown. Integrity monitoring exists for exactly these states of almost-rightness. It is designed to catch the moment when continuation becomes less a technical capability than a wager.

That wager becomes visible in behaviour. A mature system, once integrity drops below a meaningful threshold, does not merely display a warning and carry on as though nothing has changed. It widens its route assumptions, climbs, slows, loiters, returns, refuses descent, abandons precision landing, simplifies manoeuvre, or otherwise reduces the demands placed on uncertain navigation. Trust, here, is earned through visible discipline. Operators learn to respect a platform not because it always presses on, but because it changes conduct when the quality of its own knowledge has worsened. An aircraft that protects itself by narrowing ambition may seem less heroic than one that completes the route on paper. In the Arctic, it is often the more serious machine.

The problem, however, is that integrity logic can create a new kind of danger if its thresholds are poorly chosen or poorly understood. Set them too conservatively, and the system may abandon useful missions too often, frustrate operators, shorten coverage, and tempt crews into habitual override. Set them too loosely, and the machine may preserve convenience by spending away the very margin integrity monitoring was meant to protect. This is the central design tension. It cannot be solved by technical elegance alone because it involves competing operational goods: mission completion, recoverability, asset preservation, legal responsibility, human workload, and the wider cost of unexpected loss. Thresholds are not neutral. They are a record of what the organisation is prepared to lose.

A fair limitation should be admitted. No integrity monitor can see perfectly into every failure mode. Sensors may fail in correlated ways. Models of uncertainty may be wrong. Thresholds may rest on assumptions that weather, wear or unusual operational geometry quietly violate. A system can be too eager to protect itself or too slow to recognise that its navigation quality has already decayed. Operators may misread warnings. Logs may not reveal enough about why the threshold was crossed. Integrity monitoring is not omniscience. It is a structured attempt to keep uncertainty from acquiring the privileges of certainty.

But that limitation strengthens rather than weakens the case for threshold discipline. Precisely because the system cannot know everything, it needs clear rules for what to do when enough doubt accumulates. Those rules should be explicit, testable and phase-sensitive. They should not ask the machine to make a philosophical judgement about truth in the abstract. They should ask whether the current navigation state remains sufficient for the current action set. Can the aircraft continue low over ambiguous terrain? Can it still enter a narrow recovery corridor? Can it accept this route deviation? Can it complete one more leg and still return with a margin that means anything? The best thresholds are not ornate. They are operationally legible.

This legibility matters because human operators live downstream of the machine's decisions. A system that drops below integrity limits and changes behaviour without making the reason intelligible invites override and resentment. A system that reports its integrity state clearly and couples that state to understandable behaviour stands a better chance of being trusted even when it becomes inconvenient. This is where integrity monitoring meets doctrine. People must know, before launch, what kinds of navigation doubt will force a return, a climb, a route simplification or a landing refusal. Otherwise, the machine's caution will look like a mood rather than a method.

That doctrine is also how organisations reveal what they truly value. Some will prioritise asset recovery over all else. Some will accept more navigation uncertainty to preserve persistence or collection objectives. Some will draw different threshold lines for different airframes, areas, weather regimes or mission types. None of these choices is automatically right or wrong. What matters is whether they are admitted as choices and then enforced consistently. The Arctic punishes institutions that pretend a technical threshold relieved them of moral responsibility. It did not. It merely formalised the responsibility in a more inspectable form.

The quiet motif here is the red line. Pilots, engineers and operators all know the comfort and menace of it. A red line suggests clarity: beyond here lies a different category of consequence. Yet every red line in a real system is the product of debate, evidence, compromise and appetite for risk. Integrity thresholds work in the same way. They look crisp on the interface because someone has done the untidy work of deciding where caution must become command. The line feels objective. The judgement behind it is human.

That is why control limits under degradation should be read as a test of institutional maturity. Immature programmes tend to admire navigation performance and treat integrity logic as an accessory, useful mainly for compliance or post-hoc explanation. Mature programmes reverse the emphasis. They understand that performance matters only while the system knows when it has stopped deserving its own confidence. They test the thresholds under representative conditions. They watch how crews react to them. They refine the line where the aircraft should cease doing the elegant thing and do the safe thing instead. They know that a platform's honesty about uncertainty is as important as its raw ability to estimate position.

The unresolved question comes at the exact moment Arctic operations most dislike: the route still appears manageable, the display remains tidy, the aircraft has not yet done anything obviously wrong, and integrity logic begins withdrawing permission. At that point, what should govern the next move - the apparent smoothness of the track, the operator's hope that the uncertainty is tolerable, or the threshold that was meant to encode, in advance, how much doubt this mission could afford before control itself had to change character?

Chapter 6

Communications Loss and Link Discipline

What does it mean to call a vehicle autonomous if the moment it matters most is the moment it can no longer be reliably reached? The question sounds like philosophy until a sortie begins to hinge on something brutally concrete: a control link fading behind terrain, a video feed arriving just late enough to flatter the operator, a return command sent into silence, or a stream of telemetry that stays alive while its timing has already become too poor to support confident intervention. The bars remain green.

That is the paradox of communications in the Arctic. Distance makes unmanned systems attractive because they can extend presence where crews, ships and aircraft cannot linger easily. Yet the same geography that creates the need for them also thins the links through which they are supervised, corrected and understood. Terrain interrupts line of sight. Sea state and ship motion alter geometry. Atmosphere behaves capriciously. High-latitude coverage can be uneven in ways that look minor on a planning slide and decisive in flight. A platform may still move beautifully while the chain connecting it to human authority becomes intermittent, delayed or brittle. The signal is part of the weather.

This matters because communications are often described too narrowly, as though they were a utility available in the background rather than a governing condition of operations. A system has a link, loses a link, or regains a link. The real picture is harsher and more granular. Different tasks tolerate different communications quality. Supervising a route is not the same as hand-flying a gusty recovery. Watching a payload feed is not the same as making a split-second intervention. Moving a map update is not the same as preserving accountable records after an incident. The more one looks closely, the less helpful it becomes to speak simply of "connectivity". What matters is which authority depends on which link, with what timing, and under what assumptions about interruption.

The central tension lies there. Designers and operators want persistence, reach, low crew exposure and rich data return. Those aims push systems outward across larger areas and into more demanding conditions. At the same time, safe control depends on communications qualities that become harder to guarantee as range, terrain complexity, weather and emissions constraints grow. The result is not merely inconvenience. It is a contest between ambition and legibility. A machine that can act far away is only truly useful if its distant actions remain governable, recoverable and intelligible when the link degrades rather than vanishes cleanly.

That is why the stakes extend beyond whether a video feed freezes or a command packet drops. Communications shape the practical meaning of control itself. If latency rises unpredictably, a nominally available link may still be unfit for piloting. If bandwidth narrows, a high-resolution payload stream may survive while the telemetry needed for diagnosis becomes starved. If transmissions must be restricted for operational reasons, the system's autonomy logic suddenly has to carry more of the burden that constant supervision once absorbed. If a vehicle goes lost-link, the next question is no longer whether it can fly home in theory, but whether its emergency behaviour fits the geometry, weather and uncertainty of the place where the silence began. A good emergency profile feels boring. A bad one becomes a story.

The Arctic makes these questions sharper because it offers so few forgiving alternatives. In more forgiving theatres, a temporary drop in connectivity may be absorbed by dense infrastructure, nearby alternates, familiar terrain or easy recovery. In the High North, a lapse in communication can quickly alter the whole mission's character. A diverted aircraft may be physically intact and still operationally expensive if no one can reach it easily. A payload record buffered on board may remain useful only if it can later be retrieved intact and matched to trustworthy metadata. A supervisory team may retain situational awareness in broad outline while losing the precise timing required for safe intervention. Trust, here, becomes visible in behaviour: who keeps flying, who simplifies, who aborts, who accepts autonomy's next move, and who no longer believes the machine and the link are telling the same story.

The path ahead follows that story from the shape of the Arctic itself into the rules by which communications are turned into safe authority. It begins with link geometry and propagation, where the environment starts deciding where the aircraft can be heard before doctrine ever does. It then passes into latency and bandwidth, where the distinction between seeing, supervising and controlling becomes operationally decisive, before moving into lost-link behaviour, where recoverability stops being a software flourish and starts looking like emergency design. From there the argument turns to restrained transmission and the price paid when coordination must coexist with emissions discipline, before settling on the custody of information once links become intermittent enough that what is saved, dropped, compressed or delayed begins to shape intelligence value and accountability alike.

A useful way to hold these problems together is to think of a rope between climbers in poor visibility. A rope does not remove danger. It changes the form of dependence. Its usefulness depends on tension, slack, timing, terrain and what each person assumes the other can feel through it. Too much slack and warning comes late. Too much tension and movement becomes clumsy. A frayed rope is not the same thing as no rope at all; it may be more deceptive. Communications in Arctic autonomy work in much the same register. The link is not merely a channel. It is the medium through which permission, correction, explanation and recovery travel.

That is why link discipline belongs at the heart of autonomous design rather than at its margins. A platform that communicates lavishly in easy conditions may still be poorly governed if its operators do not know which functions remain safe under degraded latency or intermittent loss. A platform that appears resilient after link interruption may still be dangerously brittle if its lost-link logic assumes a neat route home through a messy environment. A system that stores data efficiently may still fail the mission if the wrong things are preserved and the right things sacrificed. Communications do not simply support the operation. They define which parts of the operation remain humanly meaningful at distance.

The hardest question, then, is not whether the system can keep flying when the link begins to fray. Many can. The sharper question is whether the aircraft, its operators and its rules all agree on what kinds of authority survive each degradation step - and whether that agreement remains intact when the signal bars still look reassuring enough to tempt everyone into believing that connection and control are still the same thing.

Arctic link geometry and propagation

In July 2000, when two Russian mini-submarines descended to inspect the wreck of HMS *Edinburgh* in the Barents Sea, the operation depended not only on mechanical endurance and underwater navigation, but on a far less romantic truth: remote systems are only as useful as the channels through which they remain intelligible to those directing them. The mission belonged to salvage, not to Arctic drone operations, yet the lesson carries cleanly across domains. Distance is not empty. It has shape.

That shape matters because communications in the Arctic are so often spoken of as though they were an invisible utility, available until abruptly absent. The reality is stranger and more physical. A radio link is not simply "up" or "down". It bends around terrain badly or well. It travels cleanly over one water state and erratically over another. It behaves differently when the platform is low over broken ice, skimming a cluttered littoral, tucked behind a ridge, or flying just high enough to recover a line of sight that did not exist a minute earlier. A route can look aerodynamically sensible and communicatively foolish. The map is not the link.

This is why line-of-sight deserves to be treated as geometry rather than convenience. Show the problem plainly and it becomes operational at once. A vehicle launched from a sheltered coastal site climbs out over pale water and ice under apparently manageable conditions. The control team has strong signal at departure. Then the aircraft slips behind local topography, drops lower to pursue a task, or moves into a sector where the surface and air begin shaping propagation differently from what the neat planning line implied. Name the governing issue and the argument sharpens: communication space is not identical to flight space. Explain it and the implication becomes severe. Where the aircraft can go and where it can still be governed are often different shapes overlaid on the same map.

Terrain is the most obvious author of this mismatch. Arctic landscapes are frequently imagined as open and flat, but much operationally important ground is not. Fjords, ridgelines, coastal escarpments, broken islands, ice pressure features and even modest rises can produce sudden shadowing for line-of-sight links. The difficulty is not merely that mountains block signals - everyone knows that. It is that the geometry of shadow can change rapidly with altitude, platform attitude and mission choice. A route that remains communicatively healthy at one altitude may become uncertain a little lower, even though the aircraft feels physically safe and the task demands descent. Connectivity thus becomes a three-dimensional constraint. The link rides on the terrain no less than the aircraft does.

Sea state complicates the matter in a subtler and often neglected way. Over maritime Arctic routes, operators are dealing not with a stable reflecting plane but with a moving, irregular one. Surface roughness, spray, wave motion and the geometry of low-altitude flight over water can alter the behaviour of reflected and direct signal paths in ways that matter to control quality. Show a vessel-based operation in cold seas and the practical burden becomes clear. The aircraft is moving. The ship is moving. The sea is moving. The antennas are moving. The link therefore exists inside a continuously changing relationship of height, angle and interference. This is not exotic physics so much as repeated inconvenience elevated into operational consequence. A communications plan that treats the sea as a blank surface is often already too optimistic.

Atmospheric conditions add a third layer to this geometry, one that tends to be simplified away until a sortie proves otherwise. Arctic air is not neutral. Cold surface layers, inversions, moisture, fog, snow and ice crystals all affect how radio energy travels and how consistently it does so. Sometimes the atmosphere helps in unexpected ways. Sometimes it degrades the predictability of the link just when predictability matters more than raw reach. This is one reason the Arctic can feel treacherous even to experienced operators. Two routes that look similar in distance and profile may behave differently because the air itself is not behaving the same way. Communications, in other words, are part of the weather not merely in the poetic sense, but in the practical one.

That phrase deserves to be taken literally. If weather shapes lift, visibility and icing, it also shapes the availability and quality of remote authority. A drone that remains inside aerodynamic limits may still drift outside communicative prudence because the atmosphere has begun charging for range or low altitude in a way yesterday's sortie did not. The most serious programmes therefore stop treating propagation as a technical annex and start treating it as a planning input. Trust becomes visible here. It appears when a route is altered before launch because the likely link geometry is poor, when a lower-altitude task is declined because communications would become too fragile, or when a team accepts a less efficient corridor because it offers stronger and more predictable contact with the aircraft. Control is being preserved through geography.

A useful historical parallel comes from maritime and polar communications practice more broadly, where high-latitude operations have long exposed uneven coverage and awkward propagation. Satellite systems that perform comfortably elsewhere can present different practical value as latitude, orbital geometry and local horizon constraints begin to matter. Surface and airborne links that seem straightforward in a temperate test range become more conditional where terrain, weather and sparse infrastructure intersect. The enduring lesson is not that the Arctic is uniquely unknowable.

It is that high latitude turns communications planning into an exercise in humility. Availability on paper and availability in use are not the same fact.

A second example can be drawn from shipboard and coastal unmanned operations publicly discussed by research and patrol organisations working in northern waters. Even when platforms are capable and crews experienced, routes are often built around communicative plausibility as much as around sensor coverage or endurance. Aircraft may be kept higher than the payload team would prefer, flown along less direct legs, or held inside sectors that preserve better link behaviour relative to the host vessel or control point. The revealing point is not that connectivity matters - everyone says that. It is that connectivity quietly redraws the mission. Propagation becomes route structure.

This has important consequences for what operators mean by range. A vehicle's nominal range is an incomplete measure if the link cannot be sustained, interpreted and acted upon across that same space with enough consistency for safe authority. Too many discussions of autonomy speak as though the aircraft's fuel, battery or endurance defined how far the mission can reach. In practice, the communication envelope often trims that ambition first. The aircraft may be able to travel farther than the operator can supervise meaningfully, or farther than the data link can sustain without dangerous latency and dropout. Distance then becomes deceptive. The machine is not really as far as its map position suggests. It is as far as its last trustworthy connection allows it to remain governed.

This is one reason the Arctic rewards altitude in such a paradoxical way. Climbing often improves line-of-sight and may simplify propagation geometry. It can also reduce the quality of the mission for tasks that depend on low-level observation, close inspection, surface interaction or constrained recovery. The result is a familiar hard trade-off. To communicate better, the aircraft may have to fly in a way that sees less or acts less precisely. To perform the task more directly, it may have to descend into poorer link conditions. A wide-area feed may continue to reassure the control team right up until the moment the aircraft enters a phase where command timing and link quality matter more than video richness. The contradiction is not incidental. It is part of the operational design.

This is where the concept of the battlespace becomes useful even outside explicitly adversarial scenarios. Connectivity is shaped by terrain and weather, but it is also shaped by where an operation chooses to stand in relation to them. A mission flying a long coastal arc, threading between islands, working beneath ridgelines or staying low to preserve observational value is making communicative commitments whether it says so aloud or not. In more overtly contested circumstances, those commitments become sharper because emissions themselves may carry costs. But even without deliberate interference, the basic fact remains: the environment is already contesting the link. The Arctic behaves like a quiet adversary to any plan that assumes communication is merely a background service.

A fair limitation must be admitted. Arctic propagation is not always hostile, and not every mission suffers equally. There are routes over water or ice where line-of-sight can be sustained well. There are infrastructure-rich locations, high enough platforms and carefully positioned relays that soften many of the worst penalties. Some communications architectures are markedly better suited to high-latitude work than others. It would be lazy to suggest that every northern sortie is haunted equally by link fragility. Sometimes the geometry is generous.

But the limitation does not rescue complacency. It reinforces the central point that link design must be specific to place, profile and purpose. Good connectivity in one sector says little about the next ridge, the next lower approach, the next change in sea state, or the next weather layer. The reconciliation is straightforward and demanding: connectivity should be treated as locally earned rather than globally assumed. A route is communicatively good because its geometry, atmosphere, surface and control arrangement support it under the actual conditions of the mission, not because the platform generally has a powerful radio or an attractive coverage diagram.

This is why propagation belongs inside mission design rather than bolted onto it afterwards. The aircraft's path, height, timing and task structure should all be shaped by where communications will remain meaningful rather than merely nominal. A platform entering a sector with marginal geometry may need a different altitude band, a different loiter plan or a different expectation of how much human intervention remains plausible. A vessel-based operation may need to account not just for the aircraft's position, but for the ship's future position and motion relative to it. A coastal launch site may offer excellent departure connectivity and poor recovery connectivity once the aircraft returns lower and the weather has shifted. These are not engineering curiosities. They are route truths.

The quiet motif here is the horizon. For pilots, it is a visual reference. For communicators, it is also a radio fact. The horizon determines where sight ends and where many links begin to struggle, bend or vanish. In the Arctic, the horizon is rarely simple. Terrain lifts into it. Sea state corrupts it. weather thickens it. Satellite geometry complicates it. A sortie that ignores the horizon as a communications boundary is often just delaying the moment when the aircraft will discover that distance and command have parted company.

That is why connectivity should be read not as a feature of the platform alone, but as an interaction among platform, environment and mission geometry. A well-designed system can still be badly routed. A strong radio can still be poorly placed. A healthy feed can still mislead the team about how much authority remains available once timing and path structure deteriorate. The serious measure of communication quality is not whether data is flowing at all. It is whether the data and control relationship still fits the part of the mission the aircraft is currently attempting.

The unresolved question, then, is not whether the vehicle can still transmit as it moves farther into the Arctic scene. It may well do so. The harder question is whether the shape of the link - across ridge, sea state, cold air and shifting altitude - still supports the kind of control, supervision and recovery the mission is counting on. When that answer changes, the aircraft has not simply flown into a different place. It has flown into a different regime of authority.

Latency, bandwidth, and control authority

In July 1969, when NASA guided Apollo 11 towards the Moon, everyone involved understood a simple and brutal fact: there are tasks a machine may perform under delay and tasks for which delay becomes part of the hazard itself. The mission belonged to crewed spaceflight, not Arctic drones, yet the lesson is exact. Distance does not merely weaken contact. It changes the meaning of control. The picture can arrive late.

That truth is often hidden by modern communications language. People speak of a link as though its adequacy could be read from a single comforting sign: high bandwidth, clear video, a stable telemetry panel, a generous data rate. But bandwidth is only one property of a link, and often not the one that decides whether an aircraft remains safely governable. A system may stream broad, handsome imagery while latency has already become too erratic for precise intervention. It may deliver rich payload data while command acknowledgements lag just enough to make a gusty recovery or obstacle-sensitive manoeuvre unsafe. More data is not the same thing as timely authority.

This is why control authority must be divided rather than spoken of in the singular. Piloting, supervision and payload control are not one activity performed through the same communications pipe. They are different relationships to time. Show the distinction plainly and the operational stakes come into view. Piloting asks for low delay, predictability and continuity because it lives in the loop of action and correction. Supervision can tolerate more delay because it often concerns route approval, health monitoring, mode changes or mission framing rather than split-second control. Payload control sits somewhere in between and varies by task: a broad-area survey can absorb delay that would be intolerable in a close inspection or target-tracking sequence. Name the issue properly and the argument sharpens: link quality must be matched to the kind of authority being exercised. Explain it and the implication becomes severe. A link good enough to watch is not automatically good enough to steer.

The Arctic makes this harder because it produces exactly the sort of variable, uneven delay that encourages overconfidence. In benign settings, latency can be modest and stable enough that operators stop thinking about it consciously. In northern operations, geometry, propagation, atmospheric conditions, relay structure and intermittent coverage can produce spikes, stalls and partial dropouts that arrive not as permanent failure but as instability. That instability is worse than a clean loss in some respects, because it invites continued use. The operator still has contact. The feed still moves. The control system still answers, until the answers arrive out of phase with the environment they are supposed to command. A delayed correction is a different kind of mistake.

This is where wide-area video can become deceitful. Show a surveillance or reconnaissance feed with decent image quality and broad situational context, and it is easy to believe that the link is healthy in the ways that matter most. The scene appears continuous. The aircraft appears responsive. The human mind naturally treats visual continuity as evidence of control continuity. Then the mission descends into a more dynamic phase: perhaps a close pass, perhaps a low-level turn in wind, perhaps a landing where drift, sink rate and gust response need rapid interpretation. Suddenly the same feed that reassured the operator becomes part of the problem. Its smoothness masked delay. The payload channel looked prosperous while the command loop had already become poor. That is why safe control cannot be inferred from impressive video alone.

A useful historical example lies in remotely piloted and teleoperated systems beyond aviation, where the distinction between supervisory control and direct manual control has long been recognised. Space robotics managed by NASA offers one of the clearest public demonstrations. Operators do not hand-fly a Mars rover in any meaningful real-time sense because the communication delay makes such authority absurd. Instead, they work through supervisory logic, bounded commands and carefully structured autonomy. The relevance to Arctic systems is not that polar aircraft suffer planetary delay. They do not. It is that the mission must be honest about which form of authority the link can actually support. Once that honesty is lost, operators begin trying to exercise precise control through a channel fit only for supervision.

A second example comes from unmanned aviation and maritime systems used in remote and shipborne environments, where operators have repeatedly learned that strong bandwidth does not rescue weak timing. Publicly described operations often reveal this pattern: an aircraft or vessel can return abundant sensor data over a decent link and still become difficult to control manually when delay grows uneven near recovery, especially if moving platforms, wind or close geometry enter the picture. The significance is not merely technical. It is doctrinal. Systems that blur supervision and piloting under one generic idea of "remote control" are often the ones most likely to discover the difference in the worst phase of the mission.

Latency itself should therefore be treated less as a number than as a relationship between task and tolerance. A delay that is harmless when approving a waypoint may be dangerous when avoiding an obstacle. A spike that is barely noticeable during transit may be fatal to confidence during landing. A small dropout in a payload stream may be tolerable in surveillance and intolerable in a hand-flown final approach. This is why engineers and operators who talk only about average latency miss the Arctic problem. The environment often punishes spikes, jitter and inconsistency more than it punishes modest but stable delay. A system can be designed around known slowness. It struggles more with uncertainty about when the slowness will arrive.

Bandwidth complicates the matter because it invites a false hierarchy. We are trained by consumer technology to think of higher bandwidth as improvement by default. In remote aircraft operations, greater bandwidth can certainly help. Richer payloads, stronger telemetry, clearer imagery and better post-mission records all benefit. But bandwidth also competes with other link needs, and it can seduce programmes into treating the payload stream as the centre of communicative value. That is a mistake.

An aircraft that sends beautiful wide-area video while command and state channels degrade into intermittent uncertainty is not well connected in any operationally serious sense. The right question is not how much data the link carries. It is whether the right data arrives with the timing required by the kind of authority the mission presently needs.

This becomes sharper when one notices that different mission phases demand different link priorities. High-quality payload transmission may matter most during broad collection. Navigation and health telemetry may matter more as confidence begins to narrow. Control acknowledgement and low-latency command paths may matter most in recovery. A mature system therefore treats communications not as a fixed pipe but as a shifting hierarchy of needs. What deserves bandwidth at altitude and range may not deserve it near the deck in wind. What looks like useful video during transit may become dispensable if preserving low-delay control or essential aircraft state becomes more important. Link discipline begins when the system knows what to sacrifice first.

A fair limitation deserves emphasis. Not every Arctic mission requires tight piloting authority over a live link. Many can and should be flown under supervisory control, with onboard stabilisation, route execution and contingency logic carrying the fast dynamics that remote humans cannot safely manage. Under those conditions, moderate latency and variable bandwidth may be entirely acceptable. Some platforms are designed specifically around this truth, and they become safer because they do not pretend to support direct human control in phases where the link does not warrant it. This is a genuine strength, not a compromise to be apologised for.

But the counterpoint only sharpens the central argument. If a mission relies on supervisory control, then doctrine, interface design and emergency logic must make that explicit. Operators should know which tasks are meant to be commanded indirectly, which phases require stronger timing discipline, and when the aircraft should refuse fine-grained remote intervention because the link is not fit for it. The reconciliation is therefore clear enough. High bandwidth is valuable. Stable authority is more valuable. A system that downgrades payload richness to preserve command integrity may be behaving with greater seriousness than one that continues streaming lavishly while control becomes rhythmically late.

This is why control authority should be designed like a ladder rather than imagined as a single switch. At the top sits direct piloting, demanding the tightest latency and continuity. Below it sits constrained manual influence, where humans can still shape motion but with more onboard smoothing and guardrails. Below that sits supervision, where operators approve plans, modes, or objectives but do not close the inner control loop. Beneath that may sit payload-only influence or observation. Each rung asks something different of the link. A well-designed Arctic system knows when to step down the ladder before the link forces a fall. A poorly designed one lets operators keep climbing even after the rung has become too slippery to trust.

The same discipline should guide interface design. If the payload feed remains strong while command responsiveness worsens, the interface should not allow the former to masquerade as evidence of the latter. If latency spikes beyond safe manual-control tolerance, the system should express that change as a shift in authorised behaviour rather than as a minor communications note buried in telemetry. If a phase of flight demands low-delay control, the link state should be judged by that demand, not by whether some lower-priority channel is still performing nicely. This is what it means to connect communications engineering to control ethics. The machine should not flatter the operator into believing a form of authority remains available once the timing no longer justifies it.

The quiet motif here is the echo. In remote operations, a delayed or uneven control link behaves like an echo of authority rather than authority itself. The operator acts; the machine replies from a moment already passing away. Sometimes the echo is close enough not to matter. Sometimes it is just far enough to distort the entire exchange. A wide-area video feed can make that echo feel smaller than it is, because the eye values continuity over timing until the environment suddenly demands timing more than continuity. The echo then reveals itself in the most expensive way: through late correction.

This has consequences for training and institutional habit as well. Operators who mostly experience good links can become culturally optimistic about manual intervention, believing they can always “take over” if something begins to go wrong. Arctic conditions punish this optimism because they often degrade the link in ways that leave contact nominally alive while quietly disqualifying it from precise control. A mature programme trains people to recognise not only when the link is gone, but when it is still present and already unsuited to a certain form of authority. That is a harder lesson. It requires humility rather than just reaction.

The unresolved test comes when the aircraft is still transmitting generous video, the mission still appears recoverable, and the operator feels that intervention remains available if needed. Then the task tightens - a low turn, a descent, a gusty landing, a close pass - and the real question arrives. Is the link still supporting the authority the human believes he or she possesses, or has the operation already slipped into that dangerous interval where the image arrives richly enough to reassure, while the control loop has become only an echo of command?

Lost-link behaviours and recoverability

On 15 January 2009, US Airways Flight 1549 lost thrust in both engines after a bird strike shortly after departure from LaGuardia Airport in New York. The event was not about unmanned aircraft and not about the Arctic. Yet its lasting lesson reaches directly into both. Emergency behaviour matters most when the elegant plan has already failed and time has become geometry. A system under stress does not reveal only its hardware. It reveals what sort of future its designers thought worth preserving. The route home was gone.

That is the proper lens for lost-link behaviour. Many discussions treat it as a tidy software branch: if communication is lost, the aircraft does one of a small number of things. Return to home. Hold position. Land immediately. Continue mission. The language sounds neat because it belongs to flowcharts. Arctic operations strip away that neatness. Link loss does not occur in a vacuum. It occurs at a specific altitude, above a specific surface, with a particular state of energy, weather, navigation confidence, icing exposure, route geometry, and recovery possibility. A lost-link response that looks sensible in a test field can become dangerously brittle over broken ice, moving sea, a cluttered littoral, or a valley whose communications geometry already hinted that the "home" profile was based on assumptions no longer true. Silence changes the map.

This is why "go home" logic deserves suspicion rather than applause. Show the issue plainly and the weakness appears at once. A vehicle loses contact and turns for its launch point along a preplanned path or heading. That seems prudent, even obvious. Name the hidden premise and the whole structure starts to wobble: that the route back remains safer, better known and more recoverable than every other option. Explain why this premise often weakens in the Arctic, and the implications become severe. Conditions may have changed. Wind may now make the nominal return profile energetically worse. Ice or low cloud may have made the original approach corridor less honest than it was on departure. A vessel-based launch point may have moved. Terrain or link shadow may have caused the loss in the first place, meaning the same route back may simply retrace the geometry of failure. A brittle response is not one that fails to act. It is one that acts on stale confidence.

Recoverability is the larger concept that corrects this narrowness. A recoverable system is not merely one that can execute an emergency manoeuvre. It is one that can preserve the aircraft, preserve options, and preserve intelligibility long enough that the emergency remains governable. That sounds abstract until it is translated into visible conduct. A recoverable aircraft climbs to a safer band before deciding further. It moves towards a corridor with better navigation confidence rather than simply the shortest path home. It sheds mission ambition in an order already aligned with survival. It records enough state that, if lost, the reason will later be legible. It behaves predictably enough that operators on the ground can infer its logic even while cut off from it. The implication is straightforward and demanding. Lost-link design should be judged less like a convenience feature and more like an emergency system: by whether it produces stable, foreseeable outcomes when ordinary command authority has already failed.

This is one reason recoverability should never be collapsed into mere aircraft survival. A platform that avoids crashing but vanishes into a location no one can reach, or lands intact while discarding the data needed to understand what happened, has not fully succeeded. In Arctic operations, recovery is as much logistical and informational as it is aerodynamic. A lost-link procedure that sends an aircraft to the nearest nominal landing site may be physically conservative and strategically foolish if the site is inaccessible, politically problematic, or operationally opaque. A loiter that preserves the airframe may drain the battery below the margin needed for a controlled approach when contact returns. An automatic descent to a predesignated point may sound safe until one asks what the snow, sea state, drifted terrain or moving launch platform is like now rather than when the plan was filed. Recoverability is the ability to fail into a future the organisation can still inhabit.

A useful historical example lies in commercial aviation's treatment of emergency procedures more broadly. Mature emergency design has long understood that a checklist is not a moral triumph because it exists. It is judged by whether it creates predictable, stabilising action under stress while respecting the actual geometry of the emergency. The relevance to unmanned Arctic systems is direct. A lost-link routine should not be admired because it is automatic. It should be admired only if the automatic behaviour is legible, bounded and suited to the contexts in which link loss is likely to occur. Otherwise, automation merely turns silence into momentum.

A second example comes from the public culture around consumer and professional drones, where return-to-home functions have become almost iconic. Their success is real. Many aircraft have been saved from operator error, confusion and momentary disorientation because the platform could recover to a known point without improvisation. That deserves acknowledgment. Yet the same popular familiarity has encouraged a dangerous simplification. People begin speaking of return-to-home as though it were a general answer rather than a context-dependent one. The distinction matters enormously in the Arctic. A feature that works beautifully over a benign field can become far less reliable when home has shifted, winds have changed, the route back crosses weaker navigation space, or the original launch point is no longer the best available place to recover. Convenience has taught a habit of trust that harsher environments do not owe us.

This does not mean return profiles are inherently suspect. It means they should be designed hierarchically rather than sentimentally. "Home" may need to be interpreted less as a single coordinate than as a class of recoverable outcomes: a corridor, a loiter band, a fallback rally point, a safe altitude, or a conservative transition phase before any commitment to landing is made. Show the difference in operational terms and the design maturity becomes obvious. One aircraft loses link and immediately commits to a low, direct route towards launch origin. Another climbs, checks its navigation integrity against surviving references, evaluates energy state, chooses a safer geometry of return or diversion, and only then commits to a more specific recovery profile. The second may take longer. It is often behaving with far greater seriousness.

Predictability matters just as much as cleverness. Operators should not be forced to guess which emergency branch the aircraft has selected once contact is gone. A lost-link system earns trust when crews can answer, before launch and under stress, a plain sequence of questions. If the link drops here, what does the aircraft do first? Under what conditions will it climb, hold, return, divert or land? How long will it preserve one option before choosing another? What state does it privilege: airframe, data, energy margin, launch-point recovery, or minimum-risk landing? These are not merely training questions. They are how doctrine turns automatic behaviour into social trust. An emergency routine that surprises its own operators has already failed part of its purpose.

The Arctic makes this harder because link loss is often entangled with other degradations. Communications may weaken at the same moment that navigation confidence falls, icing begins to matter, visibility collapses, or battery margins narrow. A brittle system treats these as separate problems and applies a simple lost-link branch regardless. A resilient system understands that silence changes the weight of every other uncertainty. If GNSS confidence is also poor, a long autonomous return may deserve less trust. If the battery is cold and the wind has shifted, loiter may be less safe than early diversion. If the launch platform is moving, "home" must be dynamic or conditional. If the route to origin crosses the same terrain that caused the outage, returning on the old profile may be merely a slower way of compounding the problem.

This is where graceful failure becomes a better standard than successful heroics. Programmes are often tempted to praise systems that continue doing impressive things after losing the link. That temptation should be resisted. In harsh environments, the more impressive lost-link behaviour is often the less ambitious one: simplify, climb, narrow the action set, preserve logs, avoid novelty, and choose paths whose outcomes the organisation can explain and, if necessary, physically recover from. Graceful failure is not timidity. It is a disciplined refusal to let the absence of human authority become an invitation to improvisation by stale assumptions.

A fair limitation belongs here. Not every mission requires elaborate lost-link branching, and not every platform can carry the sensing, computation or control sophistication needed for highly adaptive recovery logic. Simpler systems can still be useful, especially when their operating area is tightly constrained, their launch geometry stable, and their environment relatively forgiving. Indeed, excessive complexity in emergency logic can itself become a liability if operators cannot predict it or if edge cases proliferate faster than testing can catch them. The criticism is real. Sophistication is not the same thing as safety.

But that limitation sharpens rather than weakens the argument. The true choice is not between crude simplicity and ornate intelligence. It is between emergency behaviour that is honestly matched to the mission space and emergency behaviour that merely sounds reassuring. A simple lost-link return can be entirely appropriate if the route structure, operating area, launch method and recovery assumptions have been tightly built around it. A more complex environment may require more layered logic because the simple answer is no longer honest. The reconciliation is therefore straightforward and stern: emergency design should be no more complicated than necessary and no simpler than the environment allows.

This also means recoverability must be tested against the actual shapes of likely failure, not against ceremonial scenarios. It is easy to demonstrate a neat return to home in open conditions with stable signals, fixed launch points and generous margins. It is much harder, and much more important, to ask what the aircraft does when the link drops behind terrain, when the platform is low over drifting ice, when the vessel has moved, when the battery is colder than expected, or when navigation uncertainty and communication loss arrive together. The best programmes are curious about these ugly combinations because they know the Arctic tends to produce them. Confidence earned only on a clean route home is not much confidence.

The informational side of lost-link behaviour deserves equal seriousness. When contact breaks, what data is preserved, prioritised or sacrificed? Does the aircraft retain enough health, position and sensor context to make later diagnosis possible if it lands out or is lost? Does it compress intelligently, or drop the wrong record because the emergency logic treats all data as equal until too late? Accountability after a lost-link event depends heavily on whether the system carried its own memory through the emergency. A platform that survives but cannot explain its path has only half survived as a programme asset.

The quiet motif here is the lifeboat. A lifeboat is not judged by how elegant it looks on calm water. It is judged by whether it launches predictably, supports the right priorities, and carries people into a survivable next condition once the ship has stopped being governable in the ordinary way. Lost-link behaviour plays the same role in autonomous systems. It is the lifeboat logic of the aircraft. That is why it must be awkwardly practical rather than theatrically capable. A beautiful emergency routine that depends on conditions already violated by the emergency is no routine at all.

The most revealing question a reader can ask of any platform, then, is not "Does it have return-to-home?" Almost all serious systems have some answer to that. The better question is "What kind of future does the aircraft create for itself when it can no longer be told what to do?" Does it seek a future that is reachable, legible and physically recoverable? Does it know when the shortest route home is not the safest one? Does it preserve options before it preserves elegance? Does it behave in ways the crew can anticipate without fantasy?

That is the unresolved test that matters in the Arctic: when the link breaks, the weather is no longer kind, the launch point may have shifted, and the route home looks simpler on the screen than it truly is in the air, what exactly does the aircraft worship - the original plan, the nearest coordinate called home, or the harder discipline of creating the least bad outcome its remaining knowledge can still honestly support?

Emissions control and operational trade-offs

In October 1962, during the Cuban Missile Crisis, a great deal of military and political power turned on a quiet discipline that rarely flatters public imagination: what to transmit, when to transmit, and what might be revealed by the act of transmitting itself. The crisis is remembered for ships, missiles and brinkmanship. It should also be remembered for a sterner communications truth. Silence is not absence. It is a choice with structure.

That truth returns with unusual force in Arctic autonomy. People often speak of remote systems as though connectivity were an uncomplicated good: more telemetry, more video, more coordination, more reassurance. Yet the same link that makes a vehicle legible also makes it more observable, more dependent on a live chain of supervision, and often more brittle when that chain must be thinned or interrupted. An aircraft or vessel can be richly connected and operationally exposed. It can also be deliberately quiet and suddenly forced to carry more of its own judgement than the designers, operators or institution had ever really tested. Every transmission spends something.

That is why emissions control should not be understood merely as a technical setting on a radio or a specialised concern for niche operations. It is a design question about authority. Show the operational tension plainly and it becomes difficult to evade. A platform carrying wide-area sensors, health telemetry and navigation updates can be supervised closely while transmitting freely. Restrict those transmissions and some combination of oversight, coordination, immediate diagnosis and flexible retasking begins to weaken. Name the issue and the argument sharpens: observability and governability are linked. Explain it and the implication follows. The more an organisation expects to restrict emissions, the more it must accept that autonomy is no longer a convenience layered on top of communications, but a substitute for some part of what communications used to provide.

The Arctic makes this substitution harsher because the environment already behaves like a quiet adversary to link reliability. Terrain, sea state, propagation oddities and distance degrade communications even before any deliberate restraint enters the picture. Emissions control therefore does not start from a clean baseline of generous connectivity. It starts from a landscape in which links are already conditional. Restricting transmissions in such a setting is not simply "doing less of the same thing". It may move the mission into a different regime altogether, one where the platform must rely more heavily on onboard judgement, preplanned logic, stored contingencies and delayed rather than continuous supervision. The burden shifts inward.

This is where the trade among observability, coordination and resilience becomes most instructive. Observability means more than being seen by an adversary or any external listener. It also means the system being visible to its own operators through telemetry, position updates, payload feeds and status reports. Coordination means the human and organisational ability to retask, deconflict, diagnose and intervene in a timely way. Resilience means the platform's ability to remain useful when one or both of those are narrowed. A richly transmitting aircraft may be highly observable and highly coordinateable while proving fragile when those conditions disappear. A quieter aircraft may preserve some forms of discretion or self-containment while shedding the very links through which humans keep its actions legible. The right balance is never automatic.

A useful historical example lies in anti-submarine and naval operations during the Cold War, where emissions decisions were often inseparable from survivability and control. The specifics varied enormously and do not need romantic embellishment here. The enduring lesson is enough. Commanders repeatedly faced the fact that transmitting could reveal, coordinate or reassure - and could also expose, constrain or invite unwanted attention. The relevant point for Arctic autonomy is not to imitate any particular naval practice. It is to recognise that communication discipline has always been a bargain between what must be known, what may safely be revealed, and what the platform can still do once contact becomes thinner by choice rather than by accident.

A second example comes from space operations, where NASA and other agencies have long designed missions around communication windows, delayed contact and periods of relative radio silence dictated by geometry rather than tactics. These systems cannot assume constant oversight. As a result, they are built to sequence activity, preserve state, tolerate deferred review and continue within carefully bounded autonomy when human contact is intermittent. The Arctic analogue is imperfect but useful. If a vehicle may need to transmit less, or may experience long interruptions in practical supervision, then mission design must begin treating communication scarcity as a normal operating condition rather than as a special failure. The platform should not become incoherent merely because the human voice is quieter for a time.

This is where emissions control reshapes autonomy design in subtle ways. A platform intended for continuous live supervision can afford habits of dependence. It may stream richly, defer decisions upward, wait for retasking, expose raw data for human interpretation, and treat lost-link logic as an exception rather than a recurring design pressure. Reduce transmissions, and those same habits become liabilities. The aircraft now needs clearer onboard prioritisation, more disciplined route logic, stronger thresholds for when it may continue or simplify, and more careful planning about what information will be stored rather than sent, compressed rather than streamed, or delayed rather than immediately acted upon. Quiet systems must think earlier.

That shift has consequences for operator psychology as well. Rich links encourage a style of command in which people feel almost physically present with the aircraft. They watch, nudge, query, correct and, in some cases, grow accustomed to intervening whenever ambiguity appears. Restrict transmissions and that habit becomes harder to sustain. The organisation must decide, often in advance, which kinds of uncertainty the machine may handle on its own and which still require communicative exposure. Trust becomes visible here. It appears when operators allow the platform to remain quieter without constantly trying to reopen the loop by habit. It disappears when teams discover that their concept of autonomy was really just continuous remote micro-management disguised by a stable link.

This is also why emissions control cannot be separated from route design and task structure. If transmissions may be reduced, routes often need to become more predictable, contingencies more explicit, and mission objectives more carefully bounded. A platform that could once be redirected on the fly may now need to carry a narrower menu of acceptable actions. A sensor collection plan that relied on real-time exploitation may need to be redesigned around onboard storage and later review. A recovery profile that once assumed constant monitoring may need to privilege safe, legible behaviour over agile adaptation. The mission becomes slightly more austere. That austerity is not a defect. It is the price of coherence under reduced communicative exposure.

A fair limitation should be acknowledged. Reduced emissions do not always improve resilience, and richly connected systems are not automatically naive. In some missions, constant communication is precisely what keeps the operation safe and politically accountable. Close supervision may be essential where airspace is crowded, recovery options poor, or legal oversight stringent. Some platforms are designed with enough onboard autonomy that moderate reduction in transmission changes less than one might expect. Others may become less, not more, resilient when emissions are restricted, because so much of their safety logic and diagnosis depends on live data exchange. Quietness is not a universal virtue.

But that limitation only sharpens the larger point. Emissions discipline should be judged by whether the whole mission architecture can remain legible and recoverable under the chosen level of transmission, not by whether silence sounds tactically impressive. A platform that reduces emissions without redesigning its supervision logic may simply become under-observed and over-trusted. A platform that transmits freely without any tolerance for later restraint may be operationally narrow in a different way. The reconciliation lies in specificity. What must be seen, by whom, at what tempo, and with what consequences if that visibility is reduced? The right answer depends on mission, environment and institutional appetite for ambiguity. What does not change is that the answer must be designed, not improvised.

One of the most revealing signs of design maturity is how a system handles partial silence. Complete silence is conceptually simple, though often operationally severe. More common are mixed regimes: low-rate telemetry, delayed payload products, burst transmission, buffered reporting, scheduled check-ins, or selective suppression of certain channels. These arrangements can be sensible. They can also create a dangerous illusion that because some information is still flowing, ordinary oversight remains intact. It may not. A low-rate position feed is not the same as supervisory control. Intermittent payload snapshots are not the same as continuous scene understanding. A delayed health log is not the same as live diagnosis. The operator who mistakes partial observability for ordinary authority is standing on very thin ice.

This is why resilience must be treated as both technical and organisational. Technical resilience concerns the aircraft's ability to continue within safe limits under reduced transmission. Organisational resilience concerns whether the people and procedures around it understand what kind of authority still exists in that regime. A team may have a highly capable platform and still perform poorly if it cannot distinguish between missions suitable for reduced emissions and missions that depend on richer communicative life. Conversely, a more modest platform may perform with impressive seriousness if its route structure, data priorities and emergency behaviours have been built honestly around the communications discipline the mission demands. Silence reveals architecture.

The quiet motif here is the lamp in a winter window. A lamp makes a dwelling easier to find and easier to live in. It also announces presence. Shutter it, and warmth remains but legibility changes for everyone inside and outside. Communications in Arctic autonomy behave in much the same way. The link lights the system for its operators, helps them coordinate around it, and reveals something of its presence to the wider environment. Dim the lamp and the occupants must already know the room well enough to keep moving without breaking what matters. Keep it too bright and other costs may enter. The question is never simply whether light is good or bad. It is whether the house was designed for the level of darkness now chosen.

This matters especially for accountability after something goes wrong. Emissions control can reduce the volume and immediacy of available data. That may be acceptable, even necessary, in some operational settings. But it means the system must be more disciplined about what it preserves onboard, what it can later explain, and how its decisions remain reconstructible once the live window into its behaviour has narrowed. A quiet platform that cannot later account for its conduct is not resilient in any full sense. It is merely obscure. Real resilience includes the ability to remain interpretable after the fact.

The unresolved question, then, is not whether transmissions should be restricted in some abstract sense. The harder question is what kind of autonomy the organisation is truly prepared to own once it chooses to make the vehicle less visible - to others, yes, but also to itself. At what point does a prudent reduction in emissions become a silent confession that the platform's supposed autonomy was really just a richly monitored dependence all along?

Data custody and degraded communications

On 28 January 1986, after the loss of the Space Shuttle *Challenger*, investigators were not dealing only with wreckage and human grief. They were also dealing with evidence: what had been recorded, what had been transmitted, what had been preserved, and what the surviving data could still honestly reveal about a system that had failed in public and in pieces. The event was not an Arctic drone incident, but its lesson belongs here with unusual force. Information does not matter only when everything goes well. It matters when continuity breaks.

That is the right starting point for data custody in severe environments. People often imagine communications as the simple movement of useful information from platform to operator. When the link is healthy, data flows. When it degrades, less data flows. That summary is technically true and operationally inadequate. In the Arctic, intermittent communications force systems to make choices about what to keep, what to send, what to compress, what to drop, what to delay, and what to protect from loss even if all else must be sacrificed. These are not clerical decisions. They determine whether the mission remains intelligible. The log survives or it does not.

This is why data custody should be treated as logistics rather than ornament. Show the practical scene and the issue sharpens immediately. A remote aircraft is collecting imagery, health telemetry, navigation traces, payload metadata, system events and maintenance-relevant status while moving through a region where the link is thin, delayed or uneven. The channel cannot carry everything at once. Name the governing problem and it becomes more exact than bandwidth shortage: information triage. Explain it and the implication is severe. If the wrong things are transmitted live, the wrong things buffered, or the wrong things discarded under pressure, the platform may still complete the sortie and return with much of its operational value already degraded. The aircraft flew. The evidence thinned.

That thinning matters because different kinds of data serve different futures. Some information is needed immediately for command and safety. Some supports intelligence value, interpretation or mission exploitation. Some is crucial later, when engineers and maintainers need to understand why a sensor drifted, why a link collapsed, why a recovery felt unstable, or why a battery behaved differently from its nominal state. Some becomes essential only after something has gone wrong, when accountability depends on reconstructing not merely where the aircraft went, but what it knew, what it was told, and how its own internal state changed along the way. A platform that carries data without a hierarchy of consequence is not neutral. It is unprepared.

This is where buffering becomes more serious than it sounds. To buffer data is not merely to store it until transmission improves. It is to decide that some piece of reality is worth preserving against the possibility of communicative scarcity. A buffered flight-control trace, a health event log, a selected payload segment, a navigation confidence record, a timestamped sequence of mode changes - each may become the difference between later understanding and permanent ambiguity. Show a lost-link or degraded-link sortie and the point is obvious enough. Operators may retain the broad outline of what happened while lacking the internal evidence needed to know whether the system behaved prudently, failed silently, or was simply operating on weaker inputs than anyone appreciated at the time. The data was present once. Custody determines whether it remains available to truth.

Compression presents a related but distinct problem. Under degraded communications, systems often compress payload products, summarise health information, or reduce metadata fidelity in order to preserve throughput. Sometimes this is sensible, even necessary. Yet compression is never innocent. It does not merely make data smaller. It changes what later questions the data can still answer. A compressed video stream may remain good enough for broad awareness and useless for subtle post-incident analysis. Summarised health telemetry may preserve trend lines while erasing transient events that a maintainer would later regard as decisive. Reduced metadata may leave imagery viewable but weaken chain-of-custody, synchronisation or confidence about where the aircraft truly was when the image was taken. Information becomes portable by becoming less exact.

That trade becomes especially sharp in Arctic operations because the value of later diagnosis is unusually high. Systems operating in cold, low-contrast, link-poor conditions often fail or degrade in ways that do not reproduce neatly indoors. A gusty recovery, intermittent connector issue, suspicious power event or drift in a sensor may only become explicable if logs, mode states, timing marks and environmental context were preserved with enough granularity during the sortie itself. Publicly described field engineering across polar and other remote operations repeatedly points to the same lesson: the environment punishes programmes that assume whatever mattered live will also be whatever matters afterwards. It often is not. The telemetry that seemed dispensable during transmission may later be what separates a diagnosed failure family from expensive guesswork.

A useful historical parallel lies in accident investigation more broadly. The value of flight data recorders and cockpit voice recorders in civil aviation is not simply that they record more. It is that they preserve the right sort of evidence for later reconstruction when human memory, fragmentary reporting and partial live knowledge are no longer enough. The relevance to Arctic autonomous systems is direct. A platform operating under intermittent communications needs its own equivalent discipline: not necessarily the same devices or formats, but the same seriousness about what must survive interruption, loss and delay so that the sortie can still be understood later. Data that cannot survive a poor link is not fully operational data.

A second example comes from remote scientific and space operations, especially those run by organisations such as NASA, where communication windows, delays and constrained downlink have long forced a hierarchy of informational value. Such missions do not simply send whatever they have. They decide what is essential for safety, what is essential for immediate mission value, what may be summarised, and what must be preserved in fuller form for later transmission or onboard retention. The lesson worth borrowing is not exotic technology. It is prioritisation under scarcity. A disciplined mission knows the difference between data needed now, data needed later, and data whose loss would cripple explanation even if the primary task appears complete.

This makes accountability an operational variable rather than a legal afterthought. When a remote system behaves unexpectedly, enters degraded mode, loses the link, diverts, lands out or suffers partial failure, the organisation's ability to act responsibly depends on what it can later show. Which commands were received? Which ones were acknowledged? What state estimates changed? What confidence dropped first? Did the aircraft choose a mode transition on the basis of decaying navigation integrity, energy state, or link conditions? Did the payload actually see the target it reported, and with what metadata confidence? These questions are not peripheral. They are the difference between learning and mythology. A programme that cannot reconstruct its own difficult sorties will soon begin explaining them through folklore rather than mechanism.

That is why data custody also affects intelligence value in the strict sense of decision usefulness. An intermittent link often tempts systems to privilege visible payload products because imagery is politically and operationally seductive. People like pictures. Pictures reassure. Pictures seem to prove presence and value. But a broad image with weak metadata, uncertain timing and thin contextual logs may be less useful than a narrower set of products tied to strong provenance and system state. Intelligence is not just seeing something. It is knowing what the thing means, when it was seen, under what confidence, by which sensor conditions, and how trustworthy the aircraft's own account of its location and attitude remained at that moment. The right frame without the right context can be operationally thin.

A fair limitation deserves emphasis. Not every mission needs the same degree of onboard data retention or the same richness of post-sortie evidence. Short, tightly bounded flights near strong infrastructure may be well served by simpler logging and thinner buffering than long-endurance, high-consequence missions in harsh and remote conditions. More retention also means more storage burden, more system complexity, more careful synchronisation and sometimes more difficulty moving data back into the enterprise later. It would be simplistic to say that more recorded data is always better.

But the limitation only sharpens the central argument. The issue is not maximal retention. It is disciplined custody. Systems should preserve what is hardest to regenerate and most critical to later truth. That usually means preserving the data whose loss would leave a gap between what the aircraft did and what the organisation could later prove, diagnose or explain. The reconciliation is therefore straightforward and demanding: store enough to reconstruct the consequential parts of the mission honestly, transmit enough to govern the live operation safely, and do not confuse visually rich output with evidential sufficiency.

This is where prioritisation logic becomes part of autonomy rather than a sidecar to it. A mature system should not wait until the link is nearly gone to decide what matters. It should know in advance which classes of information outrank others under degraded communications. Core health and state data may deserve precedence over bulk imagery. Navigation integrity and mode transitions may deserve preservation even when payload products must be reduced. Selected key frames with strong metadata may matter more than continuous but weakly contextualised streaming. Maintenance-relevant anomalies may need to be marked or protected so they are not swallowed by later compression and overwrite.

These priorities are not merely technical. They reveal what the organisation thinks it will most need when certainty is under pressure.

Overwrite policy is one of the least glamorous and most consequential parts of that design. When storage is finite, something will eventually be displaced. The question is what, and under what rules. A naive circular buffer can erase the very moments that would later explain a failure if a prolonged degraded-link event prevents timely offload. A more serious design protects segments surrounding anomalies, lost-link periods, abrupt mode changes, hard manoeuvres, suspicious power behaviour or other state transitions likely to matter later. The point is not omniscience. It is respect for the kinds of events that create enduring questions. A platform that forgets precisely when it becomes most interesting is badly designed.

The maintenance burden reveals why this matters beyond operations. In Arctic systems, many faults are intermittent, environmentally triggered and difficult to reproduce. A maintainer trying to understand a connector issue, a battery sag event, a sensor freeze, a control hesitation or a degraded navigation episode needs more than a pilot's broad impression and a handful of downlinked status messages. Good custody turns field trouble into evidence. Poor custody turns it into narrative. Narrative has its place. It is a poor substitute for preserved state history when parts are ageing, support is sparse and repeat failures are expensive.

The quiet motif here is the ledger. A ledger records movement, custody and change over time. It is dull beside the drama of the mission itself. It becomes invaluable the moment something valuable goes missing or cannot be explained. Data custody in remote autonomous operations performs the same function. It is the ledger of the aircraft's truth claims. What was observed, what was inferred, what was commanded, what was delayed, what was saved, what was compressed, what was lost. Without that ledger, the organisation may still possess data. It will possess less knowledge.

This also reframes the relationship between information and fuel. Fuel is planned, rationed, carried, monitored and prioritised because everyone accepts that movement depends on it. Information logistics deserves the same seriousness. Buffer space, downlink capacity, metadata fidelity, sync discipline, anomaly tagging, post-sortie offload, archival integrity - these are not digital housekeeping. They are the supply chain by which sensing becomes evidence, evidence becomes action, and action becomes learning. A platform that manages energy well and information badly will still fly. It will merely be less governable, less diagnosable and less accountable than it should be.

The unresolved question, then, is not whether a system can keep some data alive while the link fades. Most can. The harder question is whether the information it protects under scarcity is the information the mission, the maintainer and the later investigator will actually need once the easy live picture has vanished. When communications degrade, what does the aircraft choose to remember, and what kind of truth does that choice leave behind?

Chapter 7

Autonomy Levels and Control Limits

What does it mean to call a system autonomous when the decisive question is not what it can do, but what it should be allowed to do without fresh human permission? That is the Arctic version of the argument, and it is much harsher than the one usually heard in product demonstrations or procurement language. A machine can hold attitude, follow waypoints, classify features, adapt its path and even recover from some faults. None of that settles the real issue. The real issue is who has delegated what, under which conditions, with which limits, and with what consequences when the world becomes less legible than the software expected. Capability sounds expansive. Authority does not.

The High North strips sentiment from this distinction. Cold, whiteout, intermittent links, sparse recovery options and slow rescue convert vague talk about autonomy into a liability. A platform cannot be described honestly as "highly autonomous" if no one can say with precision whether that means stabilising itself, navigating within rules, completing a mission plan, or altering its own behaviour when conditions diverge from expectation. These are not stages on a marketing ladder. They are different forms of delegation, carrying different burdens of proof and different distributions of responsibility. Ambiguity here is expensive. When an aircraft behaves unexpectedly in severe conditions, every unclear word about autonomy becomes a future argument about blame.

That is why Arctic operations force the discussion away from intelligence in the abstract and towards control in the strict sense. The environment is not merely testing whether the machine can perform. It is testing whether the humans and institutions around it have drawn a boundary they can defend once performance becomes uneven, communication becomes delayed, and perception becomes less trustworthy than the interface suggests. An aircraft can seem marvellously independent while still depending on a hidden web of human assumptions, emergency overrides, weather margins and permissive interpretations that collapse as soon as the margin goes away. The system is not as autonomous as it seemed. The leash was only hard to see.

The central tension lies there. Organisations want reach, persistence, reduced crew exposure and faster decision cycles. Autonomy appears to offer all four. Yet the very conditions that make autonomy attractive in the Arctic also make unconstrained autonomy dangerous. Once human oversight becomes intermittent and the scene becomes ambiguous, guardrails stop being bureaucratic irritants and start becoming the architecture of honesty. Altitude floors, weather limits, icing triggers, route constraints, recovery aborts and mode reversion are not admissions of weakness. They are the terms under which delegated control remains morally and operationally defensible. A system with no visible boundaries is often not advanced. It is merely under-governed.

That matters because edge cases carry more weight in the Arctic than they do in gentler theatres. Rescue is slower. Diversion is harder. Recovery is more expensive. Minor compounding faults have more time to become major ones. Novel combinations of cold, contamination, drift, poor link quality and operator workload therefore matter disproportionately. This does not mean every rare event can be predicted or simulated. It means serious programmes must stop pretending that confidence comes from having imagined all the important failures in advance. Confidence comes from designing systems, procedures and institutions that know how to narrow authority before novelty becomes catastrophe. The most mature form of autonomy may turn out to be disciplined self-limitation.

The stakes extend beyond engineering because control architecture is also a human arrangement. Override sounds reassuring until one asks when it is possible, through what latency, under what cognitive load, and with what clarity about what the machine is actually doing. In some moments, human intervention is a real safety valve. In others, it is a comforting fiction maintained by interfaces and doctrine that lag the actual conditions of the link and the mission. Arctic operations expose this quickly. A supervisor may still nominally hold authority while lacking the timing, situational clarity or procedural room to exercise it meaningfully. Responsibility, meanwhile, remains fully present. That mismatch is one of the defining risks of autonomy under harsh conditions.

The path ahead follows this mismatch from definition into consequence. It begins by treating autonomy as delegation under constraints rather than as a single ascending scale, then moves into the safe envelopes and hard guardrails that keep machine action inside a domain humans can still answer for. From there the argument turns towards edge cases, verification and operational testing, where institutional incentives often tempt organisations to confuse smooth demonstration with earned confidence. It then reaches the harder terrain of human override, where authority must be designed rather than merely promised, before ending with the question all real failures eventually force into daylight: who is accountable once design, doctrine, operators, weather and adversarial conditions have all contributed something to the outcome?

A useful way to hold these questions together is to think of a fuse. A fuse is not an embarrassment in an electrical system. It is a declaration that some boundary matters more than uninterrupted function. Good autonomy works in much the same spirit. It is not defined by never stopping, never narrowing, never refusing, or never disappointing the operator. It is defined by knowing what must trip, what must revert, what must remain constrained, and what must be handed back to humans or to simpler behaviours before the system starts spending confidence it no longer possesses. A machine without such fuses is not freer. It is more likely to fail in a way that leaves everyone arguing after the fact about who was really in control.

That is why autonomy in the Arctic should be judged less by how many tasks the machine can perform on its own and more by how honestly the whole system states the conditions of delegation. Who decides when perception is good enough? Who decides when wind, ice, link quality or novelty has pushed the machine outside its warranted envelope? Who carries responsibility when override is possible in theory and not in practice? And what, exactly, counts as acceptable machine initiative once the environment has begun stripping away the assumptions on which that initiative was licensed in the first place?

Defining autonomy in operational terms

On 15 January 2009, US Airways Flight 1549 lost thrust in both engines after a bird strike shortly after departure from LaGuardia Airport in New York. The event belonged to crewed aviation, not Arctic drones, yet it exposed a truth that autonomous systems often allow people to forget: control is never one thing. Aircraft stability, navigation, decision-making and final authority may all be distributed differently across human beings, software and procedures, and the character of that distribution matters most when the plan has already broken. The machine kept flying.

That is the right place to begin any serious definition of autonomy. Autonomy is not a compliment, a product category or a vague sense that a platform is impressively clever. It is a practical arrangement about who decides what, under which conditions, with what permissions, and with what obligation to explain those permissions when things stop going smoothly. In the Arctic, that arrangement cannot remain fuzzy for long. Cold, low contrast, intermittent links, poor recovery options and sparse infrastructure all remove the margins that elsewhere allow vague language to survive. A system called autonomous in a brochure may prove, in reality, to be only well stabilised, lightly assisted or intermittently supervised. The word is doing too much.

This matters because the term autonomy is often used to bundle together several quite different capabilities. Show the most familiar case and the distinction becomes plain. A small aircraft holds attitude, damps gusts and flies a route cleanly. People call it autonomous. But attitude stabilisation is not the same thing as route selection. Route selection is not the same thing as mission execution against changing conditions. Mission execution is not the same thing as adaptive behaviour when the machine begins revising its own choices in response to weather, energy, perception or degraded links. Name these differences and the conceptual fog starts to clear. Explain why that matters operationally and one sees the implication immediately. If the organisation cannot state which kind of delegation is actually in play, then responsibility will become confused at the first serious anomaly.

Automated stability is the most modest and, in some ways, the most trustworthy rung in this ladder. The aircraft uses onboard control logic to keep itself level, damp oscillation, or maintain a commanded attitude more effectively than an unaided human could do continuously. That capability is real and important. It is also easy to oversell because the result looks impressive from outside. A stable machine appears self-possessed. Yet stability is not navigation, not judgement, not mission discretion. It is delegated control of motion within a tightly bounded frame. Treating it as something grander invites the first category error in autonomy debates. A platform that can keep itself upright in wind has not thereby been authorised to decide where it should go when the map, the weather and the operator's intent begin to diverge.

Assisted navigation sits farther along the chain, and with it comes a heavier burden of clarity. Here the machine may hold altitude, follow routes, maintain geofenced behaviour, or execute planned movements while a human remains meaningfully in the loop. Show a route-following aircraft in the Arctic and the temptation is to imagine that navigation has now been delegated in full. But assistance is not the same thing as independent mission choice. The system is still moving inside a human-authored world of paths, tolerances, and assumptions. Name the real capability and the discipline improves: the aircraft is assisting in the execution of a planned navigational intent. Explain the implication and it becomes a question of honesty. If the platform later deviates because the operator authorised a new route, because a geofence forced a correction, or because a supervisory command altered the mission, then the locus of agency has shifted. Calling the whole sequence "autonomous flight" conceals who actually decided.

Mission execution goes further and deserves correspondingly sharper language. A system operating at this level is not merely holding a route but carrying a structured task through several phases with limited human intervention, perhaps managing onboard resources, sequencing sensor use, or responding to expected contingencies. This is where public discussion often begins to sound grand, because the platform appears to be conducting itself with purposeful continuity. Yet even here the crucial issue remains delegation under constraints. What exactly has the machine been licensed to complete on its own? For how long? Against what environmental assumptions? With what rules for abandoning the mission, preserving the aircraft, or deferring to a human decision? Without those answers, mission execution becomes another flattering phrase in search of operational content.

Adaptive behaviour is the most rhetorically dangerous category because it is the easiest to mistake for intelligence in the broad sense. An adaptive system may revise routes, reweight sensors, shed loads, alter recovery logic or choose among contingencies according to conditions it encounters. Properly designed, such behaviour can be valuable, especially in harsh environments where rigid plans become brittle. But this is also where the moral geometry of autonomy becomes sharpest. Adaptation is not merely doing more. It is doing something different from what humans last specified because the machine has been trusted to interpret certain changes on their behalf. That trust may be well earned. It may also be weakly defined. In the Arctic, where conditions can degrade quickly and oversight may be intermittent, the distinction matters enormously. A machine allowed to adapt without a clear account of which variables matter, which thresholds govern response, and which kinds of novelty remain outside its authority is not advanced. It is under-governed.

A useful historical example comes from commercial aviation's long relationship with automation. Modern airliners have demonstrated, repeatedly, that highly capable automated systems can improve safety and precision while also creating new forms of confusion when crews do not fully understand which functions are active, which protections remain engaged, and where authority truly sits in a given moment. The relevance here is not that Arctic drones are miniature airliners. It is that ambiguity about control architecture has a long history of turning technical success into human misunderstanding. If highly trained airline crews can be caught by uncertainty about mode, authority and intervention, then programmes fielding unmanned systems into cold, low-visibility, low-margin environments have even less excuse for vague language.

A second example can be found in NASA's planetary robotics, where missions such as the Mars rovers are frequently described in public discourse as highly autonomous. That description is not wrong, but its value lies in its specificity. These systems do not enjoy unlimited freedom. They operate through carefully bounded delegation, with onboard hazard handling, motion planning and task sequencing constrained by mission design, communication delay and very clear priorities about preservation, science return and human oversight. The lesson worth borrowing is not that Arctic operations are somehow Martian. It is that serious autonomy is never a mystical property. It is a contract of permissions. The more consequential the environment, the more explicit that contract must be.

This is why Arctic conditions punish ambiguous definitions so hard. In easier settings, a team can get away with saying a platform is autonomous because nothing has yet forced them to say *how*. Links remain strong enough for frequent intervention. Weather remains forgiving enough that late corrections are possible. Recovery remains easy enough that an odd decision by the machine becomes a nuisance rather than a campaign-defining event. The High North strips away those indulgences. A gusty recovery in poor visibility, an intermittent communications window, a navigation solution narrowing under degraded reference, a sudden need to shed mission ambition to preserve battery margin - these are not moments in which rhetoric can carry the operation. Someone, or something, has to hold authority. The boundaries of that authority must already be legible.

Trust therefore has to be described in operational rather than emotional terms. Trust in autonomy is not the feeling that the system is advanced. It is the decision to let it continue acting without immediate human correction because its delegated role is well understood and its limits have been earned. That trust appears in visible conduct. An operator allows the aircraft to continue route execution through weak visibility because the machine's authority in that phase is clear. A supervisor refrains from micro-intervention because the system is authorised to manage energy or geofence response within defined bounds. A maintainer signs off a sortie not because the machine is "smart" but because its control logic remained within the role the programme had actually validated. Once trust is described this way, the value of precise definitions becomes obvious. Precision in language is a precondition for precision in responsibility.

A fair limitation should be admitted. No real system stays neatly inside one category at all times. A platform may combine automated stability with assisted navigation, limited mission execution and narrow forms of adaptation, all while human supervision rises and falls with link quality, workload and phase of flight. Any tidy taxonomy risks flattening that reality. It would be simplistic to demand that every aircraft fit one pure class and never cross the boundaries between them.

But the limitation does not weaken the need for clear definitions. It intensifies it. Precisely because real systems mix forms of delegation, organisations must be able to describe those mixes without resorting to the single flattering word autonomy as if it solved the problem. The right question is not "Is this platform autonomous?" It is "Which functions are delegated under which conditions, and what remains humanly governed when those conditions degrade?" That is a stricter, better and more Arctic question. It does not abolish complexity. It organises it.

This is also why autonomy should be discussed as a matter of control architecture rather than machine prestige. Once the environment becomes severe, the most important design questions sound almost old-fashioned. Which decisions must the machine make faster than a human can reasonably intervene? Which decisions may it make only inside predeclared guardrails? Which decisions must still remain human, even if that slows the operation? What happens when communication or perception quality means the human is nominally present but functionally late? These are not secondary implementation details. They are the structure by which a system becomes answerable to the people who field it.

The quiet motif here is the leash, though not in the crude sense often used to insult automation. A leash can be long or short, visible or subtle, fixed or dynamic. What matters is not whether it exists, but whether everyone involved understands when tension on it changes and who is meant to be holding the other end. Arctic operations make invisible leashes dangerous. A system may seem gloriously independent until weather, latency, drift or ambiguity suddenly reveal that half its decisions were never really its own and the other half should not have been. Better to describe the leash honestly than to pretend it vanished.

That honesty becomes most valuable when something goes wrong. If the aircraft fails a landing, exceeds a boundary, presses on too far, turns back too early or behaves in a way the operator did not expect, responsibility must be traceable. Was this a failure of stability logic, navigation assistance, mission execution rules or adaptive authority? Was the human meant to intervene, and could they have? Was the platform acting inside its licensed autonomy or outside it? Ambiguous definitions make these questions harder not only for investigators but for engineers and operators trying to learn. Precision in terminology is therefore not academic fastidiousness. It is part of the learning system.

The unresolved test, then, is not whether a platform can be made to look autonomous in demonstration footage. Many can. The sharper question is what exactly has been delegated to it once the Arctic removes the easy margins - and whether that delegation is defined clearly enough that, when the machine acts and the situation worsens, no one has to discover in the weather what should already have been explicit on the ground.

Safe envelopes, constraints, and guardrails

In April 2021, NASA flew *Ingenuity* in Jezero Crater, and one of the least glamorous truths behind that achievement was also one of the most important: the little aircraft did not fly because it had been granted freedom in the grand abstract. It flew because its freedom had been bound with unusual care. Altitude, timing, terrain assumptions, energy use and recovery logic were not decorative details. They were the conditions under which the machine was allowed to succeed. The rails were already there.

That is the right way to think about guardrails in Arctic autonomy. They are often described as limitations, which is true but incomplete. A limitation sounds like a subtraction from some natural ideal of unconstrained machine agency. In severe environments, the opposite is closer to the truth. Guardrails are what make delegated action meaningful at all. They define the region inside which an aircraft, vessel or ground system may still be trusted to keep going when weather worsens, perception thins, human oversight becomes intermittent and recovery options stop being forgiving. Limits keep hard missions honest.

A safe envelope is therefore not a static box around a machine's physics. It is a live agreement between environment, sensing, control logic and institutional appetite for risk. Show the problem plainly and the concept sharpens. A vehicle may be physically capable of descending lower, pushing farther into wind, remaining longer in marginal icing, or attempting one more recovery pass. That does not mean it should be authorised to do so. Name what the system needs in order to remain governable and the structure becomes clear: altitude floors, wind limits, icing triggers, route boundaries, energy margins, navigation-confidence checks, communication minima and contingency branches. Explain why these belong together and the implication is stark. A machine without guardrails is not more autonomous. It is simply allowed to spend certainty it does not possess.

This is especially true when perception is degraded. In temperate and well-supported conditions, a skilled operator may compensate for a system that wanders near the edge of its envelope. The aircraft gets one more chance. A gust is corrected late but not too late. A landing is waved off after an ugly approach and recovered on the second attempt. A sensor anomaly is managed by human judgment because the outside world still provides enough reference to catch what the machine missed. Arctic conditions take away that generosity. Low contrast, blowing snow, ice accretion, intermittent links and sparse alternates mean that a system near its control edge may no longer have a humanly affordable margin of rescue. Guardrails become the replacement for that absent grace.

Wind limits offer a particularly revealing example because wind is one of the easiest things to under-describe and one of the hardest things to negotiate safely in a remote system. People often think of wind in terms of raw tolerance: can the aircraft remain airborne, hold attitude, make headway? Those are necessary questions. They are not sufficient. The more serious issue is whether the platform can still recover predictably in gusts, crosswinds or windshear when the phase of flight becomes tight, the link becomes imperfect and the sensor picture becomes less than ideal. A machine that technically survives repeated landing attempts in strong or unstable wind may still be behaving intolerably if each attempt consumes energy, increases uncertainty and invites a harder failure on the next cycle.

This is why an automated wind limit that blocks further approaches after certain conditions are met is not a timid feature. It is an institutional refusal to let hope masquerade as procedure.

That refusal matters because repeated failed landings are seductive. After the first wave-off, the aircraft is still airborne. After the second, the runway, pad or deck is still there. Each attempt teaches a little, and that little can be mistaken for growing mastery rather than shrinking margin. A human crew with full situational awareness might sometimes justify another try. A remote or partially autonomous platform in Arctic conditions should be much less easily indulged. Show the sequence and the case for guardrails becomes nearly self-explanatory: each failed attempt may bring lower battery margin, worse icing, more confusion in the sensor stack, higher operator stress and less believable performance from the same control logic that has already failed to produce a satisfactory outcome. A hard stop is not always elegant. It can be the most honest thing in the system.

Altitude floors belong to the same family of honesty. In public discussion, altitude is often treated as a tactical choice, a matter of seeing more or less, getting closer or staying safer. In harsh environments it is also a measure of epistemic humility. The lower the aircraft flies, the less room it has to absorb navigation uncertainty, poor contrast, clutter, timing delay or sudden wind effects. A platform may be able to hold low altitude smoothly under ideal assumptions. The question is whether those assumptions remain good enough in flat light over snow, over broken sea ice, along a cluttered littoral or in a gusty corridor where communications are already fraying. A strict floor is sometimes the system admitting, in advance, that it cannot combine all its uncertainties and still deserve that proximity to the surface. The floor is not only about collision. It is about honesty.

Icing limits play a similarly severe role because they force organisations to confront what they really know about environmental degradation rather than what they wish to infer from a still-flying airframe. Ice can enter a mission gently enough that people begin bargaining with it. A little accumulation, a little drag, a little sensor disturbance, a little shift in control behaviour - none of it yet dramatic. The aircraft still responds. The route is still achievable. Yet the system is no longer the same machine that launched. Guardrails matter here because they turn accumulation into an operational threshold rather than a drifting argument. Once certain combinations of temperature, moisture, performance change or sensor degradation appear, the machine's authorised action set narrows. What matters is not whether the aircraft *can* continue for a while. What matters is whether the organisation is prepared to stand behind that continuation once the environment has begun editing the platform's character.

A useful historical example lies in conventional aviation's treatment of operating limitations more broadly. Modern aircraft are not trusted because they lack constraints. They are trusted because their constraints are explicit, trained, monitored and linked to action. Airspeed limits, flap schedules, crosswind limits, icing procedures, terrain and obstacle margins, approach minima - these do not diminish aviation's seriousness. They are aviation's seriousness. The relevance to Arctic autonomy is direct. Unmanned systems that aspire to comparable operational credibility must stop treating guardrails as awkward appendices to autonomy and start treating them as the mechanism by which autonomy becomes answerable.

A second example, again from NASA's *Ingenuity*, is useful not because Martian engineering can be copied into polar fieldwork without remainder, but because it demonstrates a certain moral style. The flights were bounded by carefully chosen conditions and phases because the vehicle's success depended on respecting what had actually been tested and understood. That is the deeper lesson. A system earns trust by staying inside the truth of its own evidence. The Arctic punishes any programme that behaves as though enthusiasm can enlarge that truth after take-off.

Still, a fair limitation must be admitted. Guardrails can be designed badly. Limits set too crudely may waste mission opportunity, frustrate crews, and produce a culture of routine override that is worse than the original problem. A system that abandons tasks too early or refuses manoeuvres it could in fact handle may become operationally timid and strategically disappointing. Overly rigid constraints can also interact badly, producing unexpected dead ends in software logic or forcing awkward transitions at precisely the wrong time. It would be naive to claim that more guardrails automatically mean more safety.

But that limitation only clarifies the real standard. Good guardrails are not maximal restrictions. They are restrictions tuned to the actual quality of the system's knowledge, the reality of its recovery options and the consequences of getting one more decision wrong. That tuning requires evidence, field testing and a willingness to let the environment correct theory. A wind threshold that looks sensible in simulation may be poorly matched to real ship motion, gust structure or sensor lag. An altitude floor that feels conservative on paper may prove barely sufficient once terrain or sea clutter are considered. An icing trigger that sounds scientifically neat may be operationally meaningless if no fielded sensor or operator can judge it in time. Constraint-based control is therefore not the art of saying no. It is the art of saying no at the point where continued yes would become organisationally dishonest.

This is why contingency triggers matter so much. A contingency trigger is not merely a warning. It is the point at which the machine's authorised future changes. That shift may be small or dramatic: climb, simplify the route, abandon the landing, stop the payload task, return along a safer corridor, hold until a timer expires, or divert to a better recovery geometry. The essential thing is that the trigger links degraded condition to changed authority. Without that link, alerts remain informational ornaments. With it, they become part of the control architecture. The machine's freedom contracts because the basis for that freedom has weakened. This is exactly what serious autonomy should look like under stress.

Human oversight makes guardrails even more important, not less. There is a persistent fantasy that a human in the loop can always provide the final safety net if automation strays near the edge. Arctic operations repeatedly punish that fantasy. Human oversight may be delayed by latency, narrowed by a poor sensor picture, weakened by workload, or simply too far away from the event to judge its texture. The more intermittent the oversight, the more the system must carry its own discipline. Guardrails are how prior human judgment survives into moments when live human intervention has become too late, too confused or too expensive. They are frozen supervision.

The quiet motif here is the handrail. A handrail does not walk the stairs for anyone. It does not remove darkness or guarantee elegance. It simply ensures that when footing becomes uncertain, there remains something already placed in the environment to stop uncertainty becoming a fall. Safe envelopes and contingency limits do the same for autonomous systems. They are rarely admired in sunny conditions. They become priceless when perception degrades and the machine begins to consume the very margin the operator had assumed was still intact.

This perspective also changes how one should read impressive demonstrations. A platform that never trips its limits may look stronger than one that does. Sometimes the reverse is true. The stronger system may be the one willing to stop, climb, defer, wave off or return because it has encountered a condition the organisation decided, in advance and with evidence, should narrow delegated control. A machine that appears endlessly accommodating may simply be spending risk in ways no one has yet been forced to account for. The question is not whether guardrails interrupt the mission. The question is whether they interrupt it before the mission begins lying about its own recoverability.

The unresolved test, then, comes at the moment when the aircraft is still airborne, the task still looks possible, the operator still hopes one more attempt might work, and the environment has begun to push several uncertainties together at once. Does the system still behave as though capability alone should govern what happens next, or do its guardrails finally assert the sterner truth that under Arctic conditions, the right to continue depends not on what the machine can still attempt, but on what the organisation can still responsibly defend?

Edge cases and the problem of novelty

On 28 January 1986, the Space Shuttle *Challenger* broke apart shortly after launch from Kennedy Space Center in Florida. The technical causes are well documented, but one of the deeper lessons has never lost its force: disasters often arrive not from the single failure everyone expected, but from a combination that seemed too awkward, too unlikely, or too context-bound to be granted full moral seriousness beforehand. The pieces had existed already.

That is the proper starting point for edge cases in Arctic autonomy. An edge case is not a theatrical impossibility lurking at the far end of mathematics. It is a condition, or a combination of conditions, that lives near the boundary of what the system, the operators and the institution have truly prepared to handle. In gentler environments, many such combinations remain survivable because time, visibility, recovery and human intervention all provide slack. The Arctic removes that slack with unusual efficiency. A rare interaction that would be irritating elsewhere may become decisive here because there is less room to diagnose, less space to improvise and less chance of cheap recovery once the machine begins behaving outside expectation. Rare does not mean negligible.

This is why novelty deserves to be treated as an operational burden rather than a philosophical curiosity. Show the practical sequence and the point becomes sharp. A platform enters light icing while visibility falls, a link becomes erratic, the navigation solution remains nominal but less honest than it looks, and the battery is colder than it should be after a delayed launch. None of these factors alone would necessarily force an abort. Together they alter the meaning of the mission. Name the problem properly and one sees why it resists easy planning: interacting degradation. Explain it and the implication is sobering. The machine is no longer facing one fault. It is facing a changed world in which the assumptions that once made each fault tolerable are quietly failing together.

The Arctic gives these combinations unusual leverage because rescue and recovery are slow. That changes the arithmetic of novelty. In more forgiving theatres, a novel fault sequence may still end in a safe landing, a nearby retrieval, a rapid maintenance inspection and an expensive but contained lesson. In the High North, the same sequence may end in a vehicle down on moving ice, a diversion to a site with weak access, an aircraft returning with too little evidence to diagnose what happened, or a team forced to choose between risky recovery and strategic loss. Novelty becomes expensive not simply because the system is surprised, but because the environment lengthens the consequences of surprise. Time stretches after the first mistake.

This is also why organisations are tempted to lie to themselves about edge cases. The catalogue of possible combinations is effectively open-ended. Weather, sensor state, route geometry, battery condition, software mode, operator workload, link quality and maintenance history can meet in too many arrangements for exhaustive enumeration to feel plausible. Faced with this, institutions often choose one of two bad responses. They either behave as though novelty can be defeated by ever longer lists, or they dismiss edge cases as the inevitable remainder of any advanced system and therefore not worth disciplined attention. Both responses are evasions. One mistakes bureaucracy for understanding. The other mistakes humility for licence.

A better approach is what might be called edge case budgeting. The phrase matters because it shifts the conversation from fantasy control to constrained stewardship. A budget does not imply that every future expense can be predicted exactly. It implies that uncertainty must still be priced, bounded and planned for. Show the idea in operational terms and it becomes practical rather than metaphorical. A programme cannot list every way in which icing, low contrast, intermittent communications and inertial drift may combine. It can, however, decide how much compound uncertainty the system is allowed to absorb before its authority narrows, how much novelty must trigger simplification rather than continued ambition, and what reserves of energy, route margin, altitude, communications or recovery space must be preserved precisely because the unexpected will not arrive singly. The implication is powerful. Edge case budgeting does not promise omniscience. It refuses to let ignorance pretend to be free.

This way of thinking is especially useful because it changes how trust is described. Trust in an Arctic autonomous system should not mean the comforting belief that the machine has been tested against everything important. That standard is impossible and, worse, dishonest. Trust should instead mean that the machine and the organisation around it preserve enough margin that novelty is less likely to turn immediately into catastrophe. It appears in visible behaviours. Routes are shorter than maximum reach. Recovery profiles keep altitude or energy in hand. Perception confidence is allowed to narrow authority early. Maintenance narratives are treated as evidence of emerging interaction, not as anecdotal noise. The system behaves as though the unknown has a claim on its reserves. That is not timidity. It is disciplined respect for how harsh environments amplify compounding failure.

A useful historical example lies in the 1988 loss of Iran Air Flight 655, though the operational context was utterly different from Arctic unmanned systems. The tragedy emerged from a chain in which stress, interpretation, system presentation and human judgement interacted disastrously. Its relevance here is not political but structural. Catastrophic outcomes often depend on interaction effects rather than on one glaring error alone. The important lesson is that organisations under pressure can become dangerously willing to flatten ambiguity into a single story, especially when several small uncertainties arrive together. Arctic autonomy faces a similar risk when low-level anomalies are interpreted in isolation rather than as parts of an emerging pattern. Novelty rarely announces itself with a label.

A second example comes from aviation automation incidents more broadly, including those in which crews found themselves confronting unexpected combinations of degraded data, mode confusion and environmental stress. The details differ from case to case, but the pattern is stable enough to matter. Systems tend to fail hardest at the seams between categories: when navigation degradation becomes a communication problem, when a sensor issue becomes an energy issue because the recovery takes longer, when weather turns a software mode transition into a human-factors failure. This is precisely the kind of seam-rich environment the Arctic creates. Publicly described experience with unmanned systems in severe conditions repeatedly points to the same truth: the mission begins in one category and ends in three.

That is why edge cases should not be imagined as freak episodes existing far away from normal operations. More often they are ordinary elements arriving in uncooperative company. A gust is normal. A delayed link is normal. Reduced contrast is normal. A slight calibration drift is normal. A marginal battery response in the cold is normal. The edge case is their meeting. This matters because organisations are much better at testing elements than meetings. Bench work, subsystem trials and orderly simulations tend to isolate variables for good reasons. Arctic reality recombines them with little regard for those reasons. The problem of novelty is therefore not only technical surprise. It is a mismatch between how systems are usually validated and how the field actually presents difficulty.

Simulation plays a role here, but only an honest one. Simulations can expose logical weakness, mode transitions, timing problems and the broad consequences of degraded inputs. They are invaluable for rehearsing combinations that would be unsafe or impractical to test repeatedly in live conditions. Yet simulation also flatters the imagination that built it. What it does not represent, it does not discipline. That is why field experience in severe environments remains irreplaceable. The point is not that field trials reveal every unknown. They reveal which unknowns the programme was not even culturally prepared to respect. A brittle system does not merely fail in the field. It often fails to look surprising in exactly the right way.

A fair limitation should be stated clearly. Edge case budgeting can become empty rhetoric if it is used as a grand phrase for vague caution. One can say "we must allow for the unexpected" without changing routes, reserves, training, thresholds or testing discipline in any meaningful way. Worse, organisations can use the language of uncertainty to evade accountability, as though anything unusual automatically lay outside reasonable anticipation and therefore outside responsibility. That move should be resisted. The existence of novelty does not excuse lazy design. It sharpens the obligation to decide, in advance, what kinds of uncertainty the system must absorb gracefully and what kinds must force retreat.

This is why budgeting is the right word rather than merely awareness. A budget asks what is being spent, what must be reserved and what losses cannot be afforded. In Arctic operations, the relevant currencies are not only energy and time but confidence, communications quality, navigation integrity, recovery plausibility and operator attention. When several of these begin degrading at once, the budget is being consumed. A mature programme does not wait for one metric to hit zero before admitting trouble. It treats compounding strain as expenditure from a shared account. That mental model is far better suited to harsh environments than the more comfortable fiction that each subsystem can be judged on its own line item until a formal limit is crossed.

The quiet motif here is the crack in lake ice. No one sensible expects to map every crack in advance or to assume that a frozen surface is either perfectly solid or wholly broken. What matters is learning how strain propagates, where weakness concentrates, how much weight can still be borne, and when one more step stops being prudent. Edge cases in autonomy work much the same way. Novelty is rarely a meteor from outside the system. More often it is a crack propagating through several ordinary layers at once until the surface no longer deserves the confidence it did a moment earlier. The danger lies in pretending the crack remains local when it has already begun to travel.

This perspective also improves the conversation about responsibility. When a novel combination defeats a platform, the first institutional temptation is often to ask whether the exact sequence had ever been predicted. That is a poor standard. A better question is whether the programme had budgeted realistically for interaction, preserved sufficient margins, and designed the control logic to narrow authority as compounding uncertainty grew. If the answer is no, then the failure is not redeemed by the fact that the exact pattern was new. Novelty in harsh environments is not an alibi. It is the terrain in which serious systems are expected to move.

The reader should therefore come away with a sterner but more useful way of speaking. It is not necessary, or even possible, to enumerate every dangerous combination the Arctic may present. It is necessary to ask what combinations the system can survive without improvisation, what reserves it must keep because the unknown will consume some of them, and what triggers tell the organisation that novelty is no longer something to be managed inside the mission but something forcing the mission to become smaller, simpler and more recoverable. That is disciplined uncertainty, not fatalism.

The unresolved test comes when the aircraft is still airborne, the mission still appears broadly feasible, and no single parameter looks dramatic enough to justify alarm. A little icing. A little link instability. A little navigation doubt. A little drift in the sensor picture. A little more energy consumed than planned. At what point does the organisation stop asking whether any one of these is disqualifying and start recognising that novelty is already spending from a common budget the mission may not be rich enough to afford?

Verification, validation, and operational testing

On 4 June 1996, Ariane 5 Flight 501 lifted off from Kourou and broke apart less than a minute later after software and assumptions carried over from Ariane 4 met a flight regime they had not been built to survive. The catastrophe belonged to launch vehicles, not Arctic drones, but its lesson has never been merely about code. A system can be thoroughly tested against the wrong reality. It can pass what was asked and fail what was actually needed. The test was honest. The world was harsher.

That is the right place to begin when speaking about verification, validation and operational testing in severe environments. These terms are often recited with bureaucratic confidence, as though they naturally produce trust. They do not. Verification asks whether the system was built as specified. Validation asks whether the specified thing was actually the right one to build. Operational testing asks something still more unforgiving: whether the machine, the software, the procedures and the people remain credible once the field begins rearranging the assumptions under which all earlier success seemed reasonable. Arctic operations make these distinctions unusually sharp because the environment is so good at turning neat compliance into brittle performance. A platform can satisfy the document and still fail the season.

This is why confidence should be treated as earned under representative conditions rather than awarded by accumulated technical ceremony. Show the ordinary sequence and the danger becomes obvious. A system behaves beautifully in simulation, repeats the same success on a range, passes checklist-driven qualification and then enters the Arctic, where cold soak, blowing snow, poor visibility, intermittent links, degraded references and awkward maintenance rhythms begin arriving in combinations rather than as isolated variables. Name the real problem and the argument clarifies at once: representativeness. Explain it and the implication becomes severe. A test campaign that does not expose the machine to the kinds of interactions the mission will actually depend upon is not merely incomplete. It risks teaching the organisation the wrong kind of confidence.

Verification matters because badly built systems do not become trustworthy through poetry or ambition. Interfaces must conform, control laws must execute, sensor timing must align, thresholds must trigger correctly, logs must preserve what later diagnosis needs, and safety functions must behave as designed. In that sense verification is indispensable and often unglamorous. It is where engineering discipline stops itself from drifting into impressionism. Yet verification can also flatter because it is usually conducted against declared requirements, stable procedures and repeatable conditions. A platform may therefore emerge with strong evidence that it is doing what it was told to do without any corresponding evidence that what it was told to do remains wise once the Arctic begins subtracting margin. A correct machine can still be the wrong machine.

Validation is meant to correct this by asking whether the operational concept itself deserves trust. The question is harsher than it sounds. Not "Does the aircraft follow the route?" but "Should this route-following authority exist under the conditions in which the aircraft will be used?" Not "Does the system detect icing-related degradation?" but "Is the chosen response to icing still defensible when links are intermittent, recovery options sparse and the battery colder than planned?" Validation forces programmes to stop admiring internal coherence and begin comparing that coherence to the messy external world. In severe environments, this is where many attractive ideas begin to fray. The system may be doing the right thing according to a simplification the field does not honour.

A useful historical example lies in aviation certification and accident history more broadly, where failures have often emerged not because nothing was tested, but because interactions among software, sensors, pilot expectations and operating context were insufficiently represented before service. Public inquiries into automation-related accidents have repeatedly shown the same underlying weakness: systems were proven against requirements that did not fully capture the way people, modes, environment and degraded data would meet in practice. The relevance to Arctic autonomy is direct. A drone does not need a dramatic crash to reveal the same pattern. It needs only to perform well in tidy scenarios and then become confusing, opaque or overly bold when cold, poor perception and delayed communications begin to arrive together.

A second example comes from NASA's remote and planetary robotics programmes, where simulation has long been indispensable and never sufficient. Public descriptions of rover and helicopter testing repeatedly show an institutional seriousness about the gap between the model and the world. Simulations can exercise control logic, timing, contingency handling and sensor interpretation. They can reveal fragility early and cheaply. Yet they are always haunted by what they do not represent: the texture of real dust, thermal drift, hardware wear, communication quirks, interaction with terrain that was simplified in the model, and the behaviour of people interpreting the machine under time pressure. The lesson worth borrowing is not one of exotic technology. It is epistemic modesty. Good simulation is powerful because it is treated as a disciplined approximation, not as a flattering substitute for the field.

That distinction matters because simulation is both indispensable and seductive. It allows engineers to vary conditions systematically, test dangerous or expensive contingencies, exercise rare mode transitions and observe behaviour that would be impractical to induce repeatedly in live environments. For autonomy, this is invaluable. It is often the only reasonable place to explore combinations of failure or degradation without paying in broken airframes or reckless risk. But simulation also inherits the culture and imagination of the team that built it. What is modelled receives attention. What is omitted remains silent. A simulated Arctic may contain cold, wind and sensor noise while lacking the handling errors, maintenance shortcuts, unexpected condensation paths, changing sea state, human hesitations and cumulative small indignities that so often dominate real operations. The model can be rigorous and still too clean.

This is why field trials matter, and why they must be designed to reveal brittleness rather than merely accumulate successful sorties. There is a powerful institutional temptation to use operational testing as theatre: a sequence of demonstrations chosen to prove readiness, reassure sponsors and preserve schedule. Such trials have their uses, but they are often poor at finding what really matters. A field campaign that avoids bad light, repeated handling, awkward recoveries, partial communications loss, long-duration exposure, operator workload and borderline maintenance circumstances may produce a comforting report and a brittle fleet. Show the difference and the point becomes unmistakable. One programme treats field trials as confirmation. Another treats them as confrontation. The second is the one more likely to deserve its later confidence.

Brittleness is the key word because harsh environments do not only punish raw incapacity. They punish graceful-looking systems whose performance collapses once one or two assumptions are violated at the same time. An autonomy stack that handles each known degradation independently may still behave badly when a navigation uncertainty meets a link delay during a low-contrast recovery with rising wind and a battery already below its preferred temperature. The field is full of these meetings. A serious operational test therefore tries to force the system into awkward conjunctions, not for sadistic reasons, but because conjunctions are where the mission will eventually be decided. A test that never makes the machine choose among several degrading truths has not yet asked the best question.

This is also where institutional incentives begin to matter in ways engineers often recognise and organisations often deny. Schedules reward clean progress. Procurement rewards visible milestones. Demonstrations reward success more than candour. Operators want capability sooner rather than later. Engineers want to protect hardware, budgets and credibility. All of these pressures incline a programme towards testing that confirms the broad idea rather than destabilises it. That drift is understandable. It is also dangerous. A system can accumulate a large body of apparently positive evidence while remaining poorly tested against the conditions that would actually matter in Arctic service. The absence of embarrassing outcomes then begins to masquerade as robustness. It is merely unprovoked.

A fair limitation should be stated plainly. No test campaign can reproduce the full Arctic, still less every seasonal, regional and operational variant of it. Field testing is expensive, time-consuming and itself bounded by safety, logistics and weather. Even the best trials will miss some combinations and misjudge some future novelty. It would be foolish to imply that representative testing can ever become complete testing. Programmes that chase total representativeness will often discover only delay, exhaustion and a false sense that enough volume can replace the need for disciplined judgement.

But that limitation does not excuse shallow evidence. It argues for better selection. The point is not to test everything. It is to test the combinations most likely to expose where the concept of operations is leaning on assumptions too delicate for the environment it claims to master. That means choosing trials that stress perception while links degrade, that examine recovery after repeated exposure and handling, that force autonomy to narrow authority under genuine rather than theatrical uncertainty, and that include the people, procedures and support rhythms through which the machine will actually be used. The reconciliation is simple to state and hard to live by: test enough of the right ugliness that the organisation's confidence becomes chastened rather than merely enlarged.

The same principle should shape how success is read. A successful trial is not always the one in which the aircraft completes the mission without visible drama. Sometimes the more revealing success is a conservative abort, a mode reversion that happened early enough, a lost-link routine that looked dull but legible, or a maintenance discovery that prevented the next sortie from inheriting a hidden weakness. When programmes reward only uninterrupted completion, they train everyone involved to hide or rationalise the very signals from which real reliability grows. This is one reason operational testing is inseparable from culture. The field does not merely evaluate the aircraft. It evaluates whether the institution is mature enough to learn from a machine that says "not this way".

The quiet motif here is the mirror. A good test programme is not a stage on which the system displays its best angles. It is a mirror held close enough that the organisation can no longer avoid the asymmetries it would rather call minor. Simulation offers one mirror, polished and controllable. Field trials offer another, cracked by weather, awkward handling and human timing. Both are useful. The error lies in preferring the mirror that flatters. Arctic confidence cannot be built from flattering reflections. It has to survive the mirrors that make the machine look less elegant and more true.

This is why verification, validation and operational testing are not simply technical phases on a development timeline. They are the process by which responsibility is made real. A verified system tells us the design was implemented correctly. A validated system tells us the underlying idea was worthy of trust. A properly designed operational test tells us whether that trust survives contact with the conditions and institutions in which the machine will actually be asked to carry part of human authority. Strip away the acronyms and stage gates, and the matter is almost moral in its simplicity. No one should delegate hard decisions to a platform whose confidence has been earned mainly by avoiding the kinds of situations the Arctic will eventually insist upon.

The unresolved question, then, is not whether a programme has tested its autonomy. Almost all serious programmes can say yes. The harder question is what the tests were designed to reveal: smooth compliance with prewritten expectations, or the specific points at which the machine, the procedures and the humans around it become brittle when the environment starts combining its penalties in earnest. If the answer is the first, then confidence remains a schedule artefact. If the answer is the second, the system may still be imperfect, but its imperfections are at least beginning to belong to the truth.

Human override and authority design

On 1 June 2009, Air France Flight 447 disappeared over the Atlantic after a sequence in which degraded data, changing automation behaviour and human intervention failed to assemble themselves into a recoverable whole.

The event was not Arctic and not unmanned, but its lesson travels with unusual force into both. Human override is not valuable merely because it exists. It is valuable only if it arrives in time, with enough clarity, through an authority structure that still makes sense under stress. The hand on the controls came late.

That truth is routinely softened by the language used around autonomous systems. Override is often spoken of as a reassuring backstop, a simple safety valve waiting behind the machine's confidence. If the software behaves oddly, a human takes over. If the route becomes questionable, a supervisor intervenes. If a landing deteriorates, manual authority resumes. In benign demonstrations, this sounds almost self-evident. In the Arctic it is anything but. Remote operators do not inherit instantaneous control merely because a button, mode or procedural phrase says they do. They inherit whatever authority latency, workload, visibility, communications quality, interface design and system legibility still allow them to exercise. A nominal override can be a practical fiction.

That is why authority design matters more than the raw existence of human intervention. Show the operational problem plainly and the stakes become visible at once. A platform is flying in low contrast; the link is degraded but not fully lost; the payload feed still looks useful; the aircraft begins to behave in ways the operator did not expect; and the choice arises: trust the machine, adjust the mission, or seize control. Name the real issue and the scene sharpens: transfer of authority under uncertainty. Explain it and the implication becomes severe. If the handover point, timing and decision rights were not designed for this exact kind of ambiguity, the human and the machine may both act on half-truths at once. Override then stops being a remedy and becomes another moving part in the failure.

The Arctic punishes this especially hard because it compresses several burdens into the same moment. Latency rises. Visibility degrades. Navigation confidence narrows. Energy margin may already be shrinking. Operators are often supervising more than one variable and sometimes more than one platform or payload function. Even a small delay or ambiguity that would be manageable in a temperate test range can become corrosive here because intervention arrives into a scene already stripped of generous reference. A human trying to override late is not stepping into a stable cockpit with clear cues and direct feel. The human is stepping into a mediated, partial world built from delayed data, imperfect imagery and whatever the machine has chosen to surface. Authority becomes informational before it becomes physical.

This is why human override should be treated not as a binary but as a ladder of increasingly difficult interventions. At one end lies strategic authority: setting mission intent, boundaries and abort criteria before launch. Then comes supervisory authority: approving route changes, task priorities, contingency branches and conservative mode shifts while the machine retains fine control. Harder still is tactical influence during flight, where humans shape behaviour within a moving loop. Hardest of all is direct manual takeover in a time-critical phase, when the environment is already demanding fast judgement. These are not interchangeable forms of control. A system that preserves strategic and supervisory influence may still be wholly unsuited to late manual rescue, and a programme that fails to distinguish among them will eventually discover the difference in the weather.

A useful historical example lies in commercial aviation's long and difficult education in automation mode awareness. The public record of incidents and inquiries has shown repeatedly that the handover between human and machine is not merely a mechanical event. It is cognitive and organisational. Crews do not simply "take over". They must understand what the aircraft believes, which protections remain active, what state the sensors are in, and what their own authority now means relative to the machine's prior actions. The relevance to Arctic autonomous systems is direct. A remote operator who does not know whether the aircraft is protecting itself from wind, link degradation, low-confidence navigation or energy limits cannot meaningfully override with confidence. The first task is not commanding the machine. It is understanding the machine's account of reality quickly enough to make command morally and operationally defensible.

A second example comes from NASA's remote robotics, where communication delay forces designers to accept a stern truth: some forms of direct human control are simply not plausible in the moment and must be replaced by supervised autonomy, bounded command structures and carefully designed fallback logic. The Arctic is not Mars, but the design lesson is valuable precisely because it strips away romantic language. If the link, the workload and the pace of the situation make meaningful direct override unrealistic, then safety cannot be founded on the comforting phrase "a human can always intervene". It must be founded on architecture that remains coherent when humans cannot do so in time. Good systems are honest about which rung of authority the human truly occupies.

Still, honesty about limited override does not mean surrendering human responsibility. Quite the opposite. It means relocating human authority to the points where it can still be exercised well. This is where control architecture becomes a moral choice. An organisation decides, in effect, which decisions it trusts the machine to make without fresh permission, which it reserves to supervisors, which it forces into preplanned constraints before launch, and which it treats as too time-sensitive or too ambiguous for late intervention to be responsible. These are not merely software parameters. They express a view about how much risk a human should be allowed to inherit in real time and how much should already have been prevented by route design, guardrails and degraded-mode logic. Architecture is ethics in procedural form.

Workload is one reason this moral dimension is easy to miss. People often talk about override as though operators were sitting in calm reserve, waiting for the machine to err. In reality, remote operations in harsh environments involve monitoring links, sensors, route progress, weather, energy state, payload tasking, recovery geometry and often the behaviour of supporting infrastructure or moving launch platforms. An operator already carrying this burden may reach for override not because it is clearly the best option, but because ambiguity has become intolerable. The act feels responsible. It may still be mistimed or poorly informed. This does not indict the operator. It indicts any architecture that uses the possibility of human intervention as a substitute for better prior design.

That substitution appears frequently in public discourse. A system is praised because a human is "always in the loop", as though those words alone solved the hard part. But being in the loop means very different things depending on latency, authority boundaries and informational quality. A person watching a wide-area feed with delayed telemetry is in a loop of some kind. Whether that loop supports meaningful override in a gusty landing or a low-altitude obstacle decision is another matter entirely. The Arctic exposes this by making timing precious and reference scarce. Where the environment is severe, a human may be legally or institutionally in charge while functionally too late to perform the role that phrase invites others to imagine.

A fair limitation belongs here. Not every mission needs rich human override, and not every platform becomes safer by keeping a human poised for constant intervention. In some cases, late human takeover can be more dangerous than continued machine control, particularly when the operator's picture is delayed or partial and the aircraft's internal stabilisation or contingency logic remains stronger than the person's fleeting comprehension. There are missions where the safest architecture is one in which human influence remains supervisory and bounded rather than direct and improvisational. That matters because it prevents the discussion from collapsing into a sentimental preference for human hands simply because they are human.

But the limitation only sharpens the broader argument. The real question is not whether humans or machines are generally better. It is whether the authority relationship has been designed honestly for the conditions under which the system will actually be used. A platform that claims safe manual takeover in a link-degraded Arctic recovery should have earned that claim under representative timing, imagery and workload conditions. A platform that cannot support such takeover should say so in its architecture by forcing earlier decisions, simpler contingencies and more conservative autonomous behaviour. The reconciliation is severe but useful. Override should exist where it can be exercised well, and it should not be invoked rhetorically where it cannot.

This is why timing deserves such attention. Override that comes too late is not merely less effective. It can be structurally misleading. It encourages organisations to accept a more aggressive baseline of autonomous behaviour on the assumption that humans can rescue the outliers. The rescue rarely arrives at the ideal moment. It arrives after confidence has already degraded, after the machine has already committed to a path, after the operator has already spent cognitive effort interpreting ambiguous cues, and often after the easiest alternatives have vanished. A serious control architecture therefore asks not only *can* a human intervene, but *by when must that intervention occur to remain meaningful*? Once phrased that way, many casual claims about human backup begin to look thin.

Interface design also belongs in this argument because override depends on more than bandwidth and permissions. It depends on whether the system presents enough of its own reasoning, mode state and uncertainty to make intervention intelligible. A machine that simply behaves strangely and invites takeover has already burdened the operator with diagnostic labour at the worst possible time. By contrast, a machine that clearly signals why it is narrowing authority, why it is refusing a landing, why it is abandoning a route or why it no longer trusts a sensor gives the human a basis for either assent or intervention. Good override begins with explanation. Without it, the operator is not really overriding the machine's decision. The operator is interrupting a process only partly understood.

The quiet motif here is the handrail. In a dark or icy stairwell, a handrail is valuable not because it lets a person perform miracles. It is valuable because it has already been placed where trouble is likely, and because it reduces the need for last-second heroics. Human override should work in much the same spirit. A good architecture does not rely on dramatic, perfect interventions under peak ambiguity. It creates earlier, clearer, better-placed opportunities for human judgement and leaves fewer moments in which a late grab for control has to substitute for design honesty. The handrail matters before the stumble, not only after it.

This perspective also changes how organisations should talk about failure. If an override attempt goes badly, the easy explanation is often to focus on the operator's choice or reaction time. Sometimes that is warranted. Just as often, it is too shallow. One must ask whether the operator was asked to intervene too late, through too much delay, with too little explanatory context, against a machine state already shaped by a control architecture that assumed human rescue more generously than the environment permitted. Responsibility does not disappear into the machine, but neither should it settle lazily on the last human action visible before the outcome hardened.

The most serious programmes, therefore, design authority backwards from the worst plausible moment. If the link is degraded, what remains meaningfully human? If visibility is poor, what forms of intervention still make sense? If workload spikes, which choices should already have been taken out of the operator's hands by earlier guardrails? If the machine cannot be safely overridden in the final phase, then what earlier thresholds should force a simplification, diversion or abort before late intervention becomes the only remaining hope? These are not signs of distrust in either humans or automation. They are signs that someone has accepted that control is a timed and situated thing, not a moral talisman.

The unresolved question, then, is not whether a person somewhere in the system retains formal authority when autonomy begins to fray. The harder question is whether that authority still exists in a form that can be exercised well once latency, workload and ambiguity all begin arriving at once. When the machine does something troubling in Arctic conditions, is override a real hand on the handrail, or merely a comforting story told too late about who was supposed to be in control?

Accountability when autonomy fails

On 28 January 1986, after the Space Shuttle *Challenger* broke apart shortly after launch from Kennedy Space Center in Florida, the public argument did what such arguments often do: it rushed towards a culprit. Was it the hardware, the weather, the managers, the warnings, the schedule, the culture, the decision on the morning? The honest answer was more difficult and therefore more useful. Responsibility was distributed across a chain of design assumptions, institutional pressures and human judgements that had been interacting long before the vehicle left the pad. No single word was enough. That is why the lesson still matters.

Arctic autonomy produces the same temptation towards simplification. When a remote system is lost, misbehaves, lands out, violates a boundary or produces a harmful decision, the first pressure is to find the nearest explanation with a face. Blame the operator for poor judgement. Blame the software for a bad mode transition. Blame the weather for being unusually severe. Blame the adversary, if one is present or even imagined. Blame the hardware, the link, the map, the battery, the icy sensor, the delayed recovery. Each of these may contain truth. None is likely to contain the whole truth by itself. Failure in severe environments is rarely democratic, but it is often collective.

That is why accountability deserves to be understood as a design problem before it becomes an argument. Show the familiar sequence and the point sharpens quickly. A platform loses confidence in its navigation, a link degrades, the operator intervenes late, the recovery logic chooses a poor branch, the aircraft is damaged or lost, and the post-incident conversation begins. Name the real issue and the language improves at once: attribution across interacting causes. Explain it and the implication becomes severe. If the organisation lacks a disciplined way to assign responsibility across design, doctrine, human action, environmental condition and external interference, then every failure becomes politically legible before it becomes technically legible. Learning slows. Defensiveness speeds up.

The Arctic makes this worse because it supplies convenient alibis. Weather is harsh enough to explain almost anything badly. Distance and sparse infrastructure make evidence harder to collect. Recovery may be delayed or impossible. Logs may be incomplete if the link degraded. Operators may have worked under conditions that already encouraged caution, improvisation or fatigue. Adversarial activity, where relevant, may be genuinely present or merely invoked because it sounds more serious than admitting a weak threshold or a shallow test campaign. In such settings, the easiest story often wins first. That is exactly why organisations need stronger, calmer language than they think they do. Without it, the climate itself becomes a rhetorical dumping ground.

A useful distinction is the one between cause and responsibility. Causes describe what helped produce the outcome. Responsibility asks who was obliged to anticipate, bound, test, mitigate or respond to those causes and whether they did so credibly. The distinction matters because people often collapse it under pressure. A storm may be a cause. That does not mean "the weather" is an adequate account if the platform was fielded with unrealistic sensing claims or vague icing triggers. An operator may have made the last visible mistake. That does not settle whether the interface, authority design, mode logic or latency made a good intervention practically impossible. A hostile electromagnetic environment may have degraded the link. That does not automatically absolve the communications architecture if it was brittle in ways already known or already testable. Cause is not acquittal.

This is why clear accountability improves engineering. When responsibility is attributed with discipline, the organisation can say something more useful than "multiple factors contributed". It can specify where the burden properly lay. Design teams may carry responsibility for mode logic that concealed degraded confidence. Operators may carry responsibility for ignoring established limits or inventing undocumented workarounds. Leadership may carry responsibility for fielding under schedule pressure without representative testing. Maintainers may carry responsibility if repeated anomalies were normalised instead of escalated. Adversary action may be real and decisive, yet still not the only relevant factor if the concept of operations assumed a communicative or navigational grace it had never truly earned. Precision in attribution is not moral theatre. It is how systems stop repeating themselves.

A second useful distinction is between blame and answerability. Blame looks for a person or team to absorb the emotional charge of failure. Answerability asks who must explain their design choices, procedural decisions, threshold settings, override assumptions and evidence claims in a form that others can inspect. Blame is attractive because it is fast and often institutionally satisfying. Answerability is slower and more technically fertile. Show the difference in practice and it is obvious enough. After an autonomy failure, a blame culture asks which operator clicked the wrong thing or which engineer wrote the bad code. An answerability culture asks which assumptions were embedded in the system, who approved them, how they were tested, what evidence supported them, what anomalies had previously appeared, and why the organisation believed those assumptions remained defensible in the actual environment of use. One culture protects hierarchy. The other protects learning.

A useful historical example lies in civil aviation accident investigation, where the most serious inquiries learned long ago that a single-cause morality tale rarely improves safety. Modern investigation culture, at its best, examines design, training, maintenance, operational context, regulatory assumptions and human action together without flattening them into equal guilt. The relevance to Arctic autonomy is direct. If an unmanned system is lost after a degraded navigation sequence in severe weather, it is not enough to say that the operator overrode too late or that the software made a poor choice. One has to ask what the system had been licensed to do, what the environment had already invalidated, what the organisation had treated as acceptable uncertainty, and which earlier opportunities for safer design or doctrine were left unused.

A second example comes from the public record of automated and partly autonomous systems more broadly, especially where organisations initially tried to blame one layer only to discover that responsibility was structurally distributed. Inquiries into automation-related failures have repeatedly exposed the same pattern: the event that looked like operator error also reflected weak interface design; the "unexpected" environmental condition had been known in adjacent domains; the software branch that failed had never been meaningfully tested under representative timing or degraded data; the oversight chain was formally intact and operationally hollow. The lesson worth preserving is not cynicism. It is that accountability improves when organisations stop treating distributed responsibility as a reason for vagueness and start treating it as a reason for more exact language.

That exactness matters especially when autonomy is involved because machine action can create a false sense that human responsibility has thinned. It has not. It has moved. A platform that chooses its own recovery path, narrows its own authority under degraded perception, or refuses a landing attempt is not relieving humans of responsibility. It is expressing decisions humans previously embedded in thresholds, priorities, constraints and authority architecture. When those decisions fail, accountability should reach upstream to the people and institutions that defined the machine's permissions, not stop at the machine's last visible action. The software acted. Someone licensed the action space.

Still, a fair limitation needs stating. Distributed accountability can become its own evasion if every actor uses complexity as a shield. Organisations sometimes respond to failure by saying that "the system" failed, as though design teams, operators, leadership, maintainers and mission planners were all diluted into one fog of contribution. That is not sophistication. It is administrative camouflage. If everyone is vaguely responsible, no one is usefully accountable. The answer is not to retreat into simple blame. The answer is to separate responsibilities sharply enough that different layers can be criticised without pretending only one layer mattered.

This is where doctrine plays a surprisingly large role. A programme with clear doctrine can attribute failure more honestly because it has already declared who was meant to decide what, under which conditions, and with which triggers for withdrawal, override or simplification. A programme with vague doctrine is almost fated to produce blame games because its participants will reconstruct authority after the event according to what best protects them. If an operator thought the aircraft could adapt, the designer thought the supervisor would intervene, and the supervisor thought the geofence or lost-link logic would narrow the mission automatically, then post-incident accountability begins from contradiction. The failure is not only technical. It is constitutional.

The same is true of adversary action. Where hostile interference, deception or targeting may have played a role, the temptation is often to let that fact dominate the narrative. Sometimes it should. Adversarial disruption can be decisive, and no serious analysis should sentimentalise it away. But adversary action should not become a solvent poured over every design or doctrinal weakness. A platform can be contested and still poorly designed for contestation. A link can be degraded by hostile action and still reveal a brittle communications concept. A navigation failure can involve deception and still expose weak integrity thresholds or over-trusting autonomy logic. Clear accountability means resisting both naive innocence and opportunistic externalisation.

The quiet motif here is the ledger. A serious organisation keeps a ledger not only of money but of decisions, permissions, anomalies, thresholds, waived concerns and claimed evidence. When autonomy fails, that ledger determines whether responsibility can be discussed as a sequence of inspectable commitments or only as a drama of personalities and excuses. Who accepted the weather margin? Who signed off the navigation mode? Who decided that latency still permitted override? Who accepted the test evidence as representative? Who ignored the maintenance narrative that hinted at repeated drift or intermittent link weakness? Accountability is not discovered at the accident site. Much of it is already written in the ledger beforehand.

This is why blame games are so corrosive to learning. They do not merely produce hurt feelings or political manoeuvre. They actively distort future engineering by encouraging everyone to defend local innocence rather than examine shared structure. Operators become reluctant to report uncertainty if they fear retrospective scapegoating. Engineers overstate confidence in order to avoid the appearance of fragility. Managers push for tidy narratives because untidy ones threaten schedules, budgets or prestige. Maintainers learn that intermittent anomalies are safer left vague than escalated. Under those conditions, the next failure is already being prepared. Clear accountability, by contrast, supports candour because it lets people describe causal contribution without assuming that every disclosure is a confession of sole guilt.

The reader therefore needs a sturdier vocabulary than "human error" or "bad conditions" or "software fault". Better questions sound more discriminating. Which layer failed to contain the problem? Which assumption was invalidated first? Which actor had both the responsibility and realistic ability to prevent the outcome? Which actor inherited a decision space already degraded by others? Which environmental or adversarial factors were unavoidable, and which were foreseeable burdens that the design or doctrine should have budgeted for? Such language does not make accountability gentler. It makes it truer.

The unresolved test comes after the aircraft is gone, the weather is easy to blame, the software looks opaque, the operator feels exposed, and the institution wants a narrative before it has real understanding. At that moment, will the organisation ask who can most quickly absorb the fault, or who actually authorised the chain of assumptions that let design, doctrine, environment and human action combine in the way they did? One path protects reputations briefly. The other protects the next mission.

Chapter 8

Human Operators in Degraded Visibility

What happens to human judgment when the machine still speaks clearly and the world no longer does? That is the question whiteout forces upon any serious account of Arctic autonomy. Instruments continue to update. Video continues to arrive. Checklists still exist. Roles remain assigned. Yet the ordinary external references that help a person decide whether the machine's story deserves belief begin to thin or disappear. The human is not simply supervising a system. The human is negotiating with uncertainty while some of the usual grounds for confidence have already gone missing.

That is why degraded visibility is not merely a weather problem laid over an otherwise stable operation. It changes the mental economy of the crew. White ground, weak horizon, poor contrast and intermittent cues do not just make tasks harder in a general sense. They alter the relationship between perception and trust. A pilot, operator or mission supervisor can no longer glance outside the system and gain a quick, embodied reassurance that altitude, drift, closure rate or surface condition still look right. The cross-check weakens. The mind works harder. The machine begins to occupy more of the space where judgment used to breathe.

This is the point at which human factors stop sounding like a separate chapter in a training manual and become part of reliability itself. A platform can be mechanically healthy, well maintained and logically sound, yet still become operationally fragile if the humans around it are carrying more ambiguity than the concept of operations was built to admit. Workload rises not only because there is more to do, but because certainty becomes more expensive to obtain. Trust also changes character. It is no longer the easy trust that comes from repeated, comfortable confirmation. It becomes a strained decision about how much the crew is willing to rely on instruments, video and automation when the outside world is offering little help and the cost of misreading a cue may only become visible later.

That strain matters because automation surprise is rarely born from machine malice or human ignorance alone. It grows in the gap between what the machine is doing, what the human believes it is doing, and what the environment is no longer helping either party to verify. In clear conditions, crews can often absorb minor ambiguity because the world itself keeps supplying correction. In whiteout or near-whiteout, a smooth display may conceal a troubling drift in interpretation. A system mode change that would be legible in good visual reference can feel abrupt or irrational when the scene outside offers no steadying clue. The danger is not just that the machine may be wrong. The danger is that the human may discover that wrongness too late, or may intervene on the basis of a picture already thinned by delay, workload or false confidence.

The stakes therefore reach beyond individual sorties. An organisation that treats cognitive strain as a soft issue will build brittle missions even if its sensors, airframes and software all look respectable. It will overestimate how much ambiguity a single operator can absorb, how much reassurance a degraded video feed can truly provide, and how often crews can make prudent decisions under rising uncertainty without stronger procedural support. Over time, such misjudgments do not merely increase stress. They degrade doctrine. Operators improvise private workarounds. Aborts begin to feel like embarrassment instead of evidence of discipline. Silent confusion settles into the spaces between roles. The mission may still proceed. Learning begins to rot.

That is why training under severe uncertainty deserves a different moral status from rehearsal under ideal conditions. The most useful crews are not the ones who have only practised clean mission success, but the ones who can recognise a degraded situation early enough to reduce ambition without mistaking that reduction for failure. Uneventful aborts, conservative recoveries and disciplined refusals to continue often tell more truth about readiness than elegant completions. The culture around such acts matters immensely. If the organisation quietly rewards only the fluent sortie and treats hesitation as weakness, then it will eventually produce operators who remain outwardly capable while inwardly less willing to admit what they no longer know.

Procedural design becomes the next line of seriousness. Under degraded visibility, crews need more than goodwill and general competence. They need roles that remain distinct under stress, handovers that do not smuggle assumption across the table, and communications that force uncertainty into speech before it hardens into private interpretation. A strong operating concept does not merely assign responsibilities. It matches them to human limits. Who is monitoring the aircraft state, who is questioning the sensor picture, who is protecting the recovery geometry, who is allowed to interrupt, and who is responsible for naming that the situation has become less certain than the mission plan implied? Those are not administrative details. They are the grammar of safe behaviour once the environment stops offering easy confirmation.

The path through this chapter follows that grammar from the inside out. It begins with workload, trust and cognitive strain, where whiteout changes the terms on which people rely on instruments, video and automation at all. It then moves into training shaped by ambiguity rather than by tidy performance, before arriving at the practical discipline of checklists, crew roles and spoken coordination, where the abstract problem of human limitation becomes visible in handovers, monitoring patterns and the prevention of silent failure.

A useful way to hold these questions together is to think of a fogged windscreen. The vehicle may still move, the instruments may still function, and the road may still exist exactly where it did before. But the driver's work changes at once. Small judgements become heavier. Trust in the dashboard rises, then grows uneasy. What once felt effortless now demands active interpretation. Arctic operations in degraded visibility create the same shift for human crews around autonomous systems. The hard question is not whether people remain in the loop, but whether the loop they inhabit still matches what human beings can honestly perceive, say and decide when the view begins to whiten and the machine keeps sounding calm.

Workload, trust, and cognitive strain

On 28 November 1979, Air New Zealand Flight 901 flew into the slopes of Mount Erebus on Ross Island, Antarctica. The catastrophe had several causes, and it should not be reduced to a single moral about visibility. Yet one lesson from it remains painfully clear for anyone designing or operating autonomous systems in polar conditions: a scene can remain visible while ceasing to provide reliable judgement. White ground, pale sky and weak contrast can rob the human mind of its ordinary checks long before the instruments stop speaking. The outside world vanished first.

That is why degraded visibility is more than a meteorological nuisance. In whiteout or near-whiteout, workload rises not simply because operators have more to do, but because every act of interpretation becomes more expensive. A video feed must be trusted without the usual external confirmation. A navigation display must be read against a landscape that no longer offers obvious shape or scale. A mode change that would be easy to understand in clear conditions may feel abrupt or irrational once the world outside the screen has flattened into uncertainty. The result is not merely strain. It is a shift in how trust itself is formed.

Trust, in this setting, is neither blind faith in automation nor old-fashioned confidence in human instinct. It is a continuous judgement about whether the machine's account of the world still deserves authority. Show the operator's actual task and the point becomes clear. The aircraft remains on route. The altitude looks acceptable. The payload video appears smooth. The map is calm. And yet the snow surface below offers almost no independent clue that the machine is seeing, estimating and responding as the operator assumes. Name the condition and the argument sharpens: cross-check deprivation. Explain it and the implication is severe. When the outside world stops correcting the inside picture, even accurate instruments begin to feel strangely fragile, and even fragile instruments can feel persuasive for too long.

This is one reason whiteout increases workload so efficiently. The human operator is no longer merely supervising a platform. That person is also compensating for the loss of environmental reference that would normally make supervision cheaper and faster. Small ambiguities become cognitively heavy. Is the aircraft drifting, or does the weak texture of the snow only make it look that way? Is the video feed lagging slightly, or is the platform responding late to a gust? Is the machine changing mode prudently, or has something in the navigation or control stack begun to misbehave? Such questions may sound technical, but they are lived mentally and physically as strain. Reliability, here, includes the reliability of human interpretation under pressure.

A second historical example helps show why this matters beyond polar flying itself. On 1 June 2009, Air France Flight 447 was lost over the Atlantic after unreliable airspeed data and the resulting automation changes collided with human understanding under stress. The environment was different, but the structural lesson is close to the Arctic case. People are much more vulnerable to automation surprise when they no longer possess an easy external check on what the machine is telling them. In a highly mediated environment, the system can remain busy, lawful and internally coherent while the crew's mental model begins to fall behind. Automation surprise is not magic. It is often what happens when the machine changes state faster than human beings can rebuild trust in what the machine now believes.

In Arctic remote operations, that surprise can emerge in quieter but equally corrosive forms. A route-following mode may tighten its logic because navigation confidence has shifted. A landing routine may wave off because wind estimation, surface reference or control response no longer meets internal thresholds. A payload system may continue sending broad, comforting imagery while the timing of the control loop has already become less believable than the image suggests. None of these changes need be dramatic to become dangerous. What matters is that the operator, deprived of external cross-checks, may discover the significance of the change only after beginning to intervene or after trusting the system a little too long. Surprise, in harsh environments, is often delayed comprehension.

The instinctive response is to say that better training should solve this. Training does matter, but only if it is honest about the kind of mental work degraded visibility actually creates. Rehearsing perfect missions under clean reference conditions builds fluency, not necessarily resilience. What crews need is practice in reading systems whose outputs remain plausible while their meaning is narrowing, practice in speaking uncertainty aloud before it becomes embarrassment, and practice in recognising that a calm display can still impose an overloaded mind. A disciplined abort performed early, with no visible drama, is often a stronger sign of readiness than a smooth completion carried out on increasingly doubtful interpretation. That is not a sentimental point about culture. It is a practical one about preserving judgement before it hardens into private overconfidence.

A fair limitation belongs here. Whiteout does not make competent operators helpless, nor does it reduce every instrument or video feed to guesswork. Polar aviation and field operations prove otherwise. Organisations such as the British Antarctic Survey have long operated in severe visual conditions by pairing disciplined procedure with a sober understanding of what instruments can and cannot substitute for. White environments do not make trust impossible. They make it more expensive and more procedural. That distinction matters because it prevents the argument from sliding into melodrama. The issue is not that humans fail in the cold by definition. It is that the quality of their judgement depends heavily on whether the operating concept respects the costs imposed by degraded reference.

That respect becomes visible in how roles and workload are structured. If one operator is expected to monitor vehicle health, interpret video, evaluate navigation integrity, anticipate link quality, manage contingency logic and prepare recovery in whiteout conditions, then the programme is gambling against cognition, not merely against weather. Human factors are not a separate soft layer around the machine. They are part of the machine's reliability envelope, because the system includes the person who must decide when the machine is still trustworthy enough to continue. A beautifully engineered platform paired with a badly overloaded operator is not a reliable system. It is a partial success surrounded by denial.

This is also why trust in instruments and video should be discussed as a dynamic operational state rather than as a stable attitude. In clear conditions, trust can be cheap because the world keeps confirming what the display says. In degraded visibility, trust must be earned continuously against the possibility that the machine's internal order is drifting away from the operator's own uncertain grasp of the scene. The windscreen returns here as a useful motif. When the windscreen fogs, the driver does not merely "see less". The cost of every judgement rises, and the dashboard becomes both more necessary and more suspect. Arctic autonomy places operators in exactly that sort of relationship with their displays. They need them more. They can verify them less.

A serious programme therefore asks harder questions than whether the operator is well trained or whether the automation is advanced. It asks what signs tell the crew that workload is rising faster than insight, what kinds of ambiguity should force a simplification of the mission, and whether the system makes its own state legible enough to prevent silent cognitive overload. It asks whether whiteout conditions are being treated as a sensor problem, when they are also a trust problem and a workload problem. It asks whether the operating concept assumes too much about how long a human can remain a reliable integrator of delayed, partial and sometimes contradictory cues once the world outside offers no cheap correction.

The unresolved difficulty lies exactly where the system still appears calm. The aircraft is flying, the video is moving, the instruments are alive, and nothing has yet failed in a way that would justify a clean procedural answer. Under those conditions, when does rising cognitive strain stop being a private burden carried by a diligent operator and become a system-level reliability issue that the mission should already have been designed to recognise?

Training for uncertainty and ambiguity

On 11 April 1970, after an oxygen tank exploded aboard Apollo 13, NASA's mission ceased to be an exercise in executing the intended flight and became something harder: a prolonged encounter with uncertainty in which survival depended less on rehearsed perfection than on disciplined adaptation. The famous line about a problem in space has survived because it sounds dramatic. The more useful lesson is quieter. Training mattered most not where it had polished an ideal mission, but where it had taught people how to think when the mission was no longer ideal. The script had ended.

That lesson belongs in the Arctic with unusual force. Harsh environments punish the belief that competence means repeating a clean plan fluently. Whiteout, intermittent links, changing winds, thermal penalties, weak visual references and uncertain recoveries do not usually arrive as theatrical emergencies. More often they produce ambiguity: a task that still looks possible but less honest, a machine that still appears capable but less interpretable, a crew that still has options but no longer enough confidence to treat those options as equivalent. Training that focuses mainly on ideal execution leaves people impressive in calm conditions and underprepared for the moment the world stops being neat.

That is why uncertainty itself must become a training object. Not merely bad weather, failed sensors or lost links in isolation, but the human act of recognising when evidence has thinned, when trust should narrow, and when a mission should become smaller before anyone is forced into a visibly dramatic failure. Show the operational problem and the argument sharpens quickly. A crew receives plausible video, plausible telemetry and plausible route progress, yet the surface outside has gone flat, the wind is becoming less cooperative, and the recovery geometry no longer feels as clean as it did in planning. Name the real challenge and it becomes more exact than emergency response: judgement under ambiguity. Explain it and the implication becomes severe. The best-trained crew is not the one that can perform beautifully when everything confirms the plan. It is the one that can say, early and without embarrassment, that the plan has ceased to deserve its original confidence.

This is where rehearsing contingencies becomes more important than rehearsing ideal missions. Ideal missions teach sequence, fluency, timing and ordinary coordination. All of that matters. But ideal missions also flatter the assumption that success means reaching the planned end state in the planned way. Arctic operations repeatedly undermine that assumption. Success may consist in turning back before a weak signal becomes a hard problem, accepting that a payload task must be abandoned because the aircraft's confidence is narrowing, or recovering uneventfully from a situation that never became dramatic precisely because the crew recognised it early enough. A training culture that rewards only elegant completion teaches the wrong moral. It trains people to persist when they should be simplifying.

A useful historical example lies in aviation crew training after the late twentieth-century turn towards crew resource management. That movement is often remembered in broad, almost ceremonial terms, but its deeper value was practical. It accepted that technical skill alone did not secure safe outcomes once ambiguity, degraded information and interpersonal hesitation entered the cockpit. Crews had to be trained not only to execute, but to notice, speak up, reframe and interrupt. The relevance to Arctic unmanned operations is direct. A remote pilot, mission commander, payload operator or systems supervisor may not share a cockpit, but they still need a common discipline for recognising that the machine's apparent fluency no longer matches the situation's actual honesty. Technical rehearsal without that discipline is only partial preparation.

A second example comes from NASA's long culture of anomaly response, where simulations have often been designed not to glorify dramatic rescues but to teach teams how to remain orderly once the expected path collapses. The point is not to borrow spaceflight romance for polar operations. It is to notice the training logic. Serious organisations do not only rehearse what they hope will happen. They rehearse how people will behave when information is partial, priorities must be reordered, and success begins to mean an intact system and an intelligible return rather than mission completion in the original form. That is exactly the sort of psychological and procedural conditioning Arctic autonomy demands.

The phrase disciplined decision-making matters here because uncertainty does not excuse improvisation by mood. In harsh environments, people often know early that something is becoming less right, but they do not yet possess a clean procedural reason to stop. A gusty approach feels wrong. A sensor picture feels less trustworthy. A route still fits the map, but not the crew's confidence. This is where undisciplined cultures begin bargaining with ambiguity: one more pass, one more minute, one more attempt because no single parameter has yet crossed a dramatic threshold. Training should teach the opposite instinct. It should teach crews to recognise that multiple small degradations can create a real operational change even before any single alarm tells a convenient story. That is not timidity. It is competence in a place that charges heavily for denial.

This is why an uneventful abort and recovery deserves to be trained explicitly as a success condition rather than as a disappointed second-best.

The phrase sounds simple. Its implications are not. If crews only rehearse aborts as the residue of visible trouble, then they will subconsciously learn to treat them as admissions of weakness. But if they rehearse abort and recovery as a normal expression of disciplined judgement, then the act becomes cognitively and culturally available earlier, when it is most valuable. Show the difference and it is obvious enough. One team pushes on because abandoning the mission feels like failure unless the machine is already in obvious distress. Another turns back while the aircraft is still orderly, the logs still coherent and the recovery still straightforward, because that outcome had already been trained as evidence of maturity rather than embarrassment. In the Arctic, the second team is often the one that gets to fly again tomorrow.

Training for ambiguity must therefore include emotional calibration as well as technical repetition. People need repeated experience of stopping without spectacle, speaking uncertainty without shame, and ending a sortie with less data than hoped but more truth than they would have kept by pressing on. That kind of rehearsal changes what counts as a good day. It teaches that a clean return from a narrowing situation is not a failure hidden inside professionalism. It is professionalism. Without that shift, all the rhetoric about safety margins and guardrails will remain intellectually correct and culturally weak.

A fair limitation deserves acknowledgment. Training cannot remove uncertainty, and it cannot turn every operator into a flawless judge of emerging ambiguity. Some situations will always be genuinely difficult to read. Some crews will over-learn caution and become operationally timid. Some will misread weak signals despite good rehearsal. There is also a danger that contingency training can become ritualistic, a series of familiar drills that comfort the organisation without genuinely confronting the combinations of delay, degraded visibility, workload and machine behaviour that make real Arctic ambiguity so stubborn. It would be naive to claim that more rehearsal alone solves the problem.

But that limitation strengthens the main point rather than weakening it. Because uncertainty cannot be eliminated, training must be designed to make people better at living inside it without surrendering judgement. That means not only practising named emergencies, but exposing crews to ambiguous cases where the right action is conservative before the machine is visibly in trouble. It means varying the context, not merely the alarm. It means building scenarios in which recovery, simplification and non-event are the desired outcomes. And it means examining not just whether the crew followed procedure, but whether they recognised the moment when procedure had to be invoked before pride, inertia or hope made it more expensive.

This is where training intersects with doctrine. A crew cannot be expected to make disciplined conservative decisions if the surrounding institution quietly celebrates only persistence and production. What training normalises, doctrine must protect. If people are taught to abort early but later judged harshly for returning with an incomplete task, the training was insincere. If they rehearse degraded modes but live missions still reward elegant completion above truthful simplification, the organisation is teaching in two voices at once. Arctic reliability suffers badly under such hypocrisy because the environment quickly discovers which voice people obey when the situation becomes morally untidy.

The quiet motif here is the fire drill. A good fire drill is not training for flames in their cinematic form. It is training for orderly movement before panic becomes the dominant fact of the room. The value lies in making a plain exit feel normal rather than shameful. Training for uncertainty in autonomous Arctic operations should aspire to the same plainness. The crew should not need a catastrophe before choosing the conservative route home. They should already know how that choice feels, sounds and proceeds when made well.

This also changes how one should judge expertise. Expertise is often mistaken for the ability to continue smoothly through difficulty. In severe environments, a deeper form of expertise includes the ability to detect when smooth continuation has become the dangerous performance. That detection is not mystical. It is built through repeated exposure to ambiguity, repeated practice in verbalising doubt, repeated experience of transitioning to simpler modes, and repeated reinforcement that a mission ended without drama can still be a highly successful mission. The crew learns not only what to do, but what not to insist upon.

The unresolved test arrives when nothing is obviously broken. The aircraft is still flying, the feed is still moving, the route still appears broadly viable, and the crew has only that uncomfortable sense that the mission now deserves less confidence than it did ten minutes ago. At that moment, what has the training really prepared them to do: prove their competence by pressing on, or prove it by recognising that an uneventful return may be the most disciplined success the day will offer?

Checklists, crew roles, and communications

On 27 March 1977, two Boeing 747s collided at Los Rodeos Airport in Tenerife after a chain of misunderstandings, assumptions and communications failures turned an already strained environment into catastrophe. The disaster was not about Arctic drones, yet it remains one of the clearest demonstrations in modern transport history that procedural order is not a bureaucratic afterthought. Under stress, people do not simply need skill. They need agreed structure for who says what, who checks what, and who is allowed to assume what. Silence can kill quickly.

That lesson becomes even sharper in degraded visibility. When whiteout or near-whiteout strips away external reference, crews around an autonomous platform must rely more heavily on instruments, video, telemetry and one another's spoken interpretation. The machine may still be operating within limits. The data may still be arriving. Yet the human picture becomes more fragile because confirmation from the outside world has thinned. At that point, checklists, role division and communication discipline stop being mere administrative aids. They become part of the sensing system itself. The team thinks through structure.

This is why practical crew coordination belongs inside reliability rather than beside it. Show the ordinary scene and the problem becomes obvious. A remote pilot is watching aircraft state and route. A payload operator is following the sensor feed. A mission lead is holding the broader tactical or scientific objective in mind. A maintainer or systems specialist may be listening for signs of thermal, power or link trouble. If those people are not bound by clear expectations about who monitors which variables, what counts as a reportable concern, and how handovers must be performed, the operation begins to drift towards silent failure. No one person is obviously negligent. Everyone is partially saturated. The gap opens in the seams.

Checklists matter first because stress alters memory before it alters intention. People do not usually skip steps in degraded conditions because they no longer care. They skip steps because perception narrows, timing compresses, confidence oscillates and the mind begins privileging whatever feels most immediately urgent. A well-made checklist does not insult experience. It protects it. In Arctic operations, that protection is especially important because degraded visibility pushes crews towards internal interpretation at exactly the moment when external confirmation is weakest. A checklist stabilises the order of attention: aircraft state before payload fascination, recovery readiness before one more attempt, communication clarity before quiet assumption. It is not glamorous. It is one of the few defences against a mind becoming overconfident in its own fragmentary picture.

But checklists are often misunderstood because organisations admire them most when they are short, elegant and unobtrusive. In harsh environments, the better question is whether they are *timed* correctly. A beautiful pre-launch checklist is of limited use if the operation has no disciplined pause before descent into a weak-visibility recovery. A comprehensive emergency checklist may still be poor if it arrives too late, after ambiguity has already spread through the crew. Good checklist design is therefore not only about content. It is about insertion points: where the mission is forced to stop improvising long enough to name condition, authority and next action before momentum becomes its own argument. Timing is everything.

Crew roles carry the same burden. The phrase "who monitors what" sounds almost childishly simple until one watches an operation in which no one is fully sure whether the priority has shifted from mission completion to aircraft preservation, or from payload value to link integrity, or from route execution to recovery geometry. In clear conditions, teams can sometimes muddle through such ambiguity because the world outside is still providing clues. In degraded visibility, role confusion becomes much more expensive. A payload operator who assumes the pilot has noticed drift may remain silent. A pilot who assumes the mission lead is ready to call the abort may continue longer than intended. A systems specialist may hear the first signs of an anomaly and delay speaking because the formal threshold for interruption was never made clear. Error grows socially before it appears technically.

A useful historical example lies in aviation's development of crew resource management after accidents such as Tenerife. The lasting value of that tradition was not simply better manners in the cockpit. It was the recognition that authority, monitoring and communication had to be structured deliberately if teams were to function under stress without becoming captive to assumption or hierarchy. The relevance to remote Arctic operations is direct. An operator group supervising a drone, vessel or robotic system may not share a physical cockpit, but it still needs an equivalent grammar: explicit calls, protected challenge rights, handover discipline, and clarity about which person owns the decision at each stage of the mission. Without that grammar, technology only amplifies private misunderstanding.

A second example can be drawn from public descriptions of mission control cultures in aerospace, especially NASA's, where complex operations are broken into disciplined roles not because people are incapable of broad understanding, but because broad understanding under pressure is not the same thing as accountable responsibility. Again, the lesson is not to imitate another domain theatrically. It is to notice the logic. When environments are unforgiving and evidence is partial, role clarity allows people to contribute sharply without everyone trying to do everything at once. Arctic operations reward exactly this form of restraint.

This restraint is essential because workload in degraded visibility is not merely additive. It is distortive. A pilot monitoring weak visual reference through instruments and video does not simply have one more task; that person has a changed relationship to all existing tasks. A payload feed may appear more compelling than aircraft state because it is visually rich. A minor communication delay may feel tolerable until the mission enters a phase in which timing suddenly matters more than imagery. A repeated landing cue may begin to seem reliable through familiarity when, in truth, it is only becoming more psychologically sticky. Role design counters this by ensuring that not every seduction reaches every mind equally. Someone must remain strategically detached enough to say the simple, unfashionable sentence: this no longer looks honest.

Handovers are where much of this honesty either survives or dies. People often think of handovers as informational transfers, but under stress they are also transfers of confidence. A weak handover does not merely omit facts. It carries forward untested assumptions in a smoother, more authoritative voice than they deserve. "Everything looks fine" is not a handover. Neither is a flood of raw detail without a clear statement of what remains uncertain, what changed most recently, and what should trigger interruption. In degraded visibility, handovers need to preserve not only what the team knows, but the *quality* of what it knows. Otherwise the next operator or decision-maker inherits a polished story rather than a truthful one.

This is why silent failure deserves such emphasis. Many operational failures begin not with open confusion but with unspoken divergence. One person becomes worried about navigation integrity but waits for stronger evidence. Another notices increased workload but does not want to appear flustered. A third sees the aircraft still behaving smoothly and assumes that concern elsewhere must be manageable. The machine remains airborne. The operation continues. Yet the shared mental model has already fractured. By the time the disagreement is spoken, the easiest option may have vanished. Silent failure is not dramatic. It is a social form of drift, and Arctic conditions are exceptionally good at disguising it because degraded visibility makes private doubt harder to resolve quickly against the external world.

A fair limitation deserves acknowledgement. Procedures can be designed badly. A checklist can become clutter. Rigid role division can slow adaptation. Excessive callouts can swamp rather than clarify. A crew forced to follow an over-engineered operating concept may spend so much effort performing structure that it loses the very judgement the structure was meant to protect. This is real, and it is one reason experienced teams sometimes grow suspicious of procedural reform. They have seen paper discipline mistaken for operational wisdom.

But that limitation only sharpens the standard. The question is not whether more procedure is always better. The question is whether the operating concept matches human limits under actual conditions of ambiguity, delay and stress. A good concept reduces the number of silent assumptions the mission can afford. It tells the pilot what must be said aloud, tells the payload operator when mission value no longer outranks aircraft integrity, tells the mission lead when to own the abort without waiting for unanimous discomfort, and tells everyone what happens if communications or confidence degrade mid-handover. In other words, it removes work from memory and private interpretation so that attention can be spent where judgement is still genuinely needed.

This has direct implications for how one should evaluate a supposedly mature operation. The right question is not whether the crew is experienced or whether the checklist looks professional on paper. It is whether the procedure and role structure preserve clarity when the easy external cues go away. Can one person say "stop" without social friction becoming delay? Does the team know who is monitoring aircraft state, who is interrogating the sensor picture, and who is protecting recovery margin? Are handovers short enough to be usable and explicit enough to preserve uncertainty rather than bleaching it out? Does the communications discipline reveal strain early, or merely record confusion after it has already matured? Those are the marks of a humanly realistic system.

The quiet motif here is the relay baton. In a clean race, the baton passes smoothly and almost invisibly. Under pressure, in poor footing or weak sightlines, the whole result may depend on whether the handover was anticipated, spoken and physically unmistakable. Arctic crew coordination works much the same way. Authority, concern and situational understanding must pass between people clearly enough that no one is left running hard with empty hands while believing the baton is still there.

This is why procedural design should be treated as part of the vehicle, not merely part of the organisation around it. A platform that depends on perfect, unspoken teamwork in degraded visibility is not really well designed, whatever its sensors and software may suggest. A platform paired with clear checklists, realistic role separation and disciplined communications may look less romantic because more of its intelligence is distributed into people and procedure rather than concentrated in the machine. That is not a weakness. It is often the more honest form of autonomy.

The unresolved test comes when the mission is still technically possible, the video still looks persuasive, and no single alarm has yet forced a clean decision. At that point, does the operating concept still help the crew think clearly, challenge early and transfer authority without distortion - or does it leave each person quietly carrying more of the system alone than a human being should be asked to bear in whiteout?

Chapter 9

Maintenance Cycles, Spares, and Field Repair

What actually determines whether an Arctic drone force exists on a given morning: the number of airframes on the manifest, or the number of machines that can be inspected, cleaned, warmed, connected, trusted and turned around without lying to the crews about their condition? The question sounds administrative until a sortie plan begins collapsing under battery conditioning delays, fouled sensors, damp connectors and a shelter too small for the work the weather has made necessary. Readiness is a workbench question.

That is the enduring paradox of sustainment in severe environments. Remote systems are often discussed as though availability were a simple arithmetic of procurement and fleet size. More vehicles appear to promise more coverage, more persistence, more resilience against loss or weather. Harsh conditions disobey that arithmetic. In the Arctic, a parked aircraft is not resting between missions; it is accumulating consequences. Moisture condenses. Salt lingers. Plastics and seals remember the cold. Batteries age through handling as well as use. Packing, storage and transport continue the failure process while everyone is calling the platform non-operational. The mission does not pause. It changes form.

That is why maintenance must be understood not as support to operations but as the pace-setter of operations. The machine can only fly at the speed at which it can be honestly inspected, restored and re-cleared. If that claim sounds severe, it should. Under austere conditions, sortie generation depends less on headline endurance or software sophistication than on whether the physical system can survive its own handling cycle: unpacking, warming, connecting, checking, launching, recovering, drying, cleaning, charging, storing and preparing again. One stubborn fastener, one contaminated gimbal, one suspect battery, one half-trusted connector can turn a promising daily plan into a fiction. Small parts govern big ambitions.

The central tension lies there. Organisations want persistence, tempo and geographic reach. Maintenance imposes sequence, waiting, discipline and material honesty. Procurement culture often sides instinctively with the first set of desires because vehicles and payloads are visible, countable and easy to praise. Sustainment lives in slower, duller verbs: inspect, dry, log, condition, replace, quarantine, recheck. Yet the Arctic keeps proving that these verbs decide availability more reliably than any flourish in a product sheet. A force that cannot maintain its own truth becomes larger only on paper. The fleet looks healthy right up to the moment the hidden queue of deferred cleaning, doubtful parts and improvised fixes begins deciding which aircraft are truly real.

The stakes are larger than convenience or engineering elegance. Maintenance failure in the High North is not simply a matter of broken things. It is a matter of broken claims. If inspection intervals are treated as negotiable, if cold-weather wear is assumed to behave like temperate wear, if shelters are expected to support delicate procedures they were never built to support, then the organisation starts teaching itself false confidence. Crews begin flying on parts they no longer fully trust. Operators learn to treat repeated anomalies as personality rather than evidence. Maintainers inherit tasks that should have been designed out of the system and are then blamed when those tasks become unsafe or unrepeatable in wind, cold and poor light. A mature programme understands that sustainment discipline is not caution around the edges. It is the thing preventing ambition from turning fraudulent.

That maturity shows itself most clearly in how a programme thinks about time. There is flight time, obviously, but there is also corrosive time, waiting time, thawing time, drying time, queueing time, evidence time. A platform sitting in a case, on a bench or in an unheated shelter is still moving through risk. The Arctic makes storage and handling active phases in the life of the machine. Seals stiffen. Moisture migrates. Salt spray does its quiet work. Configurations drift unless someone keeps the ledger honest. The difficult truth is that non-operational time can be just as operationally decisive as a sortie, because it shapes whether the next launch begins from a truthful condition or from a hopeful one.

The path ahead follows that truth from the narrowest chokepoints to the widest organisational consequences. It begins where sustainment always becomes visible, in the maintenance cycle that quietly governs sortie tempo, then turns to spare parts planning, where cold attrition rarely respects tidy one-part-at-a-time assumptions. From there the argument moves into shelters, tools and the hard question of what field repair can actually achieve without fantasy, before settling on storage, corrosion and moisture, where "downtime" reveals itself as another operational environment. It ends in governance, where maintenance data, defect reporting and configuration discipline decide whether the programme learns across seasons or simply accumulates the same avoidable losses in different weather.

A useful way to hold these questions together is to think of a kitchen in winter. A household may own excellent ingredients, sharp knives and a powerful cooker, yet dinner still depends on whether the sink drains, the cupboard stock is honest, the worktop is clear and the damp has not quietly ruined what looked usable yesterday. Operations in the Arctic work much the same way. The glamorous item is rarely what determines whether the system can be put back into service safely and on time. It is the invisible rhythm of preparation, cleaning, storage and replacement that keeps promise attached to hardware.

That is why sustainment deserves to be read as a test of institutional character rather than mere technical endurance. Does the organisation stock parts the way failures actually cluster, or the way spreadsheets prefer to count them? Does it build procedures around what can realistically be repaired in cold shelters, or does it romanticise field ingenuity to excuse poor design? Does it treat corrosion, condensation and packing discipline as central performance factors, or as the housekeeping of people lower down the chain? The answers to those questions tell you whether the system is meant to survive a brochure cycle or a northern season.

The harder question, then, is not how many aircraft the programme owns. It is how many machines it can still turn into truthful sorties once cold wear, glove-handling, contamination, moisture, shelter limits and spare shortages have all started writing in the margins of the plan - and whether anyone has the courage to admit that the real bottleneck is not flight at all, but the bench where the next mission either becomes credible or quietly stops existing.

Maintenance as the pace-setter

In October 1914, when Ernest Shackleton's *Endurance* entered the Weddell Sea, the expedition's fate turned not only on courage or navigation but on a harder material fact: once the environment began dictating what could be repaired, preserved and reused, ambition had to answer to maintenance. The Arctic and Antarctic have always been stern about this. Machines, stores and structures do not merely support action in the cold. They decide when action remains honest. The bench rules the map.

That truth has not softened with autonomy. If anything, it has become easier to ignore because modern unmanned systems look so self-contained. They arrive as fleets, as software stacks, as networks of sensors and batteries, as promises of persistence without fatigue. Yet sortie generation in severe cold is rarely governed first by the number of vehicles available on paper. It is governed by how many can be inspected, dried, warmed, cleaned, charged, verified and turned around without fiction. Availability begins after landing.

This is why maintenance should be treated as the operational tempo-setter rather than as a service function following behind the "real" mission. Show the practical scene and the point becomes painfully clear. Several aircraft sit technically intact after a day of flying. One needs battery conditioning before anyone can trust its power behaviour. Another has a camera window lightly contaminated by fine snow and residue from a hurried recovery. A third has a connector or seal that feels just ambiguous enough under gloved handling to require a slower inspection. A fourth may be nominally ready but cannot be cleared honestly until logs are reviewed against a suspicious vibration or control hesitation. Name the issue and the argument sharpens at once: readiness is a maintenance verdict, not a fleet inventory count. Explain it and the implication follows. A programme with many aircraft and a weak maintenance rhythm has fewer real sorties than it thinks.

Inspection intervals are where this rhythm becomes visible. In mild environments, inspection is often imagined as a periodic discipline imposed on the machine from outside. In the Arctic, inspection is closer to a conversation with the environment. Cold, wind, moisture, handling and contamination write new questions into the airframe after each mission, sometimes after each movement between shelter and launch line. The machine that looked healthy at shutdown may not deserve the same confidence after thaw, drying, packing and relaunch preparation. Inspection therefore does not merely confirm readiness. It reconstructs it. Every interval is a decision about how much uncertainty the organisation is willing to carry into the next launch.

That makes chokepoints more important than raw fleet size. A single battery warming cabinet, a single clean bench for sensor work, one overworked maintainer with the only trusted torque feel for a recurring fastener issue, one narrow shelter where condensation cannot be controlled properly, one log review process slow enough to delay release of otherwise healthy aircraft - these mundane constraints often govern sortie tempo more than the headline number of vehicles. They are not incidental inefficiencies. They are the true geometry of availability. An aircraft waiting honestly on maintenance is more operationally valuable than one launched on hope because the schedule preferred movement to truth.

Battery conditioning deserves particular attention because it flatters organisations into thinking in terms of charge when they should be thinking in terms of usable state. A battery may return from flight with enough nominal energy to appear serviceable later, yet still require warming, balancing, rest or additional scrutiny before it should be trusted again in severe conditions. Show the ordinary temptation and the danger becomes plain. Under pressure, teams want to see batteries as interchangeable units in a sortie pipeline. Arctic operations punish that simplification. Batteries carry temperature history, discharge history, handling history and health history. Treating them as generic consumables speeds the schedule while quietly degrading reliability. The relevant question is not whether a pack can be installed. It is whether the next mission deserves the behaviour the pack can now honestly promise.

Sensor cleaning creates a similar but subtler choke point. Payloads and navigation aids do not need dramatic contamination to become operationally suspect. A thin film on a lens cover, fine snow packed around a gimbal interface, residue on a protective window, moisture migrating into a cavity, salt traces from maritime exposure - these are the sort of defects programmes are tempted to wave through because the sensor still powers on and still produces output. But in low-contrast and high-uncertainty conditions, a small optical or mechanical compromise can have disproportionate consequences. Cleaning therefore becomes more than housekeeping. It is an act of restoring interpretability. If the sortie depends on trustworthy perception, then cleaning time is mission time.

A useful historical example lies in aviation's long recognition that dispatch reliability depends less on theoretical fleet strength than on the discipline of inspection, servicing and defect control. Commercial aviation did not become dependable by pretending aircraft were ready because they had landed intact. It became dependable by institutionalising the right to delay, defer and quarantine until confidence could be justified. The relevance to Arctic unmanned systems is direct. A drone fleet that launches on superficial readiness may look efficient for a while. It will eventually discover that deferred doubt compounds faster in the cold than optimism can outrun it.

A second example comes from polar and remote scientific operations, where publicly described campaigns have repeatedly shown that instrument and vehicle availability are constrained less by dramatic breakdown than by the slower attrition of icing residue, battery conditioning, sensor fouling, packaging effects and the simple scarcity of clean, warm work space. The lesson is not exotic. It is almost banal, which is why it matters. Harsh environments do not only destroy machines spectacularly. They narrow the honest rate at which those machines can be put back into service. Programmes that fail to respect this end up confusing possession with readiness.

A fair limitation should be admitted. Fleet size does matter. More vehicles can provide resilience against single-platform loss, allow staggered use, and reduce pressure on any one aircraft if the sustainment system is designed properly. Likewise, some platforms are genuinely easier to inspect, clean and turn around than others. Better enclosure design, simpler battery handling, more maintainable payload mounts and cleaner health reporting can materially increase sortie generation. It would be lazy to suggest that maintenance alone explains all operational tempo.

But that limitation only strengthens the main point. Extra vehicles help only if the maintenance system can absorb them without turning them into a queue of partially trusted machines. A better-designed platform helps only if the programme uses that advantage to preserve honest readiness rather than inflate mission promises. The reconciliation is straightforward. Fleet size creates potential. Maintenance converts potential into actual availability. In the Arctic, that conversion is the real battle.

This is where sustainment ties directly into mission planning rather than trailing behind it. A serious operator planning sorties in severe conditions does not merely ask what the aircraft can do in the air. The harder question is what the maintenance cycle can sustain on the ground without accumulating hidden doubt. How many sensor cleans can be done properly between flights? How many batteries can be conditioned without rushing? How many inspections can be carried out by people, under shelters and at temperatures that preserve the meaning of the checks? Tempo built without these answers is not tempo. It is deferred maintenance disguised as operational ambition.

The quiet motif here is the funnel. Missions enter wide at the planning stage: many aircraft, many possible tasks, many attractive routes. They narrow at the maintenance funnel, where only so many can be cleaned, conditioned, checked and cleared honestly. The funnel is not an embarrassment. It is the point where the organisation discovers how much of its declared capability is real. A mature programme studies the funnel. An immature one curses it and tries to pour faster.

This is why maintenance culture is inseparable from truth-telling. When sortie pressure rises, organisations begin sending subtle signals about what kind of doubt they are willing to ignore. A smudged sensor becomes "good enough". A battery with an untidy temperature history becomes "probably fine". A recurring minor anomaly becomes "just how this aircraft behaves". These are not harmless adjustments. They are cultural edits to the meaning of readiness. Once normalised, they change the whole pace of the force, not by increasing true availability but by lowering the threshold for self-deception. That may work for a day. The Arctic eventually presents the invoice.

The unresolved question, then, is not how many aircraft the programme owns or how impressive their endurance looks in a test report. The sharper question is how many truthful sorties the maintenance cycle can produce before inspection intervals tighten, batteries demand patient conditioning, sensors demand proper cleaning, and the workbench begins refusing to certify the fantasy pace the operations plan would prefer.

Spares forecasting under harsh attrition

In November 1915, after the *Endurance* was crushed in the Weddell Sea, Ernest Shackleton's expedition learned a lesson that belongs as much to modern maintenance planning as to polar history: remote operations are not defeated only by dramatic loss. They are narrowed, piece by piece, when the small things that keep a system whole are no longer available at the moment they are needed. Scarcity starts tiny.

That is the right way to begin thinking about spares in the Arctic. People are naturally drawn to major components because they seem decisive: the airframe, the battery, the payload, the motor, the navigation unit. Those parts matter. Yet harsh attrition usually enters the system through humbler paths. Cold stiffens seals. Moisture finds connectors. Gloves reduce dexterity and increase handling stress. Repeated packing, unpacking and cleaning abrade edges that seemed durable in temperate testing. Fine contamination settles into interfaces never meant to be treated as consumables. A sortie is rarely cancelled because the whole aircraft has suddenly ceased to exist. More often, it is delayed or degraded because a small part has failed in exactly the place where the machine cannot be honestly cleared without replacement.

This is why spares forecasting in severe environments should not begin with single parts in isolation. It should begin with the patterns by which parts fail together. Show the ordinary field sequence and the logic becomes sharp at once. A connector begins to feel less positive in engagement after repeated glove-handling. The seal around it no longer sits with the same confidence after repeated cold cycles and hurried cleaning. Moisture ingress becomes more plausible. Electrical intermittency follows. Soon the operator is not dealing with one bad connector, but with a family of related vulnerabilities: contacts, backshells, sealing surfaces, caps, retention features, cleaning materials, perhaps even mounting strain and cable routing. Name the principle and the discipline improves: failure clustering. Explain it and the implication becomes practical. Programmes that forecast spares part by part often miss the fact that harsh environments damage systems in related groups.

That is why connectors and seals make such a useful example. They seem minor enough to be forgotten in strategic planning and consequential enough to halt operations when neglected. Under repeated glove-handling, cold shrinkage, contamination and hurried assembly in poor shelter conditions, these parts do not behave like independent line items. They fail as a family. One worn seal encourages moisture. One ambiguous fit encourages rehandling. Rehandling increases wear on retention and contact surfaces. Cleaning meant to restore confidence may itself accelerate degradation if the materials or methods are not suited to the environment. What began as a spare-parts problem becomes a reliability and tempo problem. Stocking wisely, in such cases, means forecasting the failure ecology rather than only the catalogue number.

This is one of the reasons harsh attrition is often misunderstood. People hear the phrase and imagine catastrophic cold damage, dramatic breakage or spectacular wear. The more common reality is slower and more administrative. Attrition arrives through the repeated indignities of use: gloved fingers applying slightly uneven force, connectors mated in awkward posture, housings opened in poor light, snow entering work areas, salt traces surviving a rushed wipe-down, seals compressed when colder than the designer intended, parts stored slightly damp, and repeated transport vibration turning minor looseness into recurring doubt. Each event is small. The cluster is operationally large. A spares model that ignores these ordinary insults is not conservative. It is abstract.

A useful historical analogue lies in polar expeditions and remote field science, where public accounts have long shown that material endurance is rarely defeated by a single grand failure. It is worn down through repetition, handling, weather and the slow conversion of minor defects into significant limitations. That is one reason experienced polar organisations tend to develop a stern respect for unglamorous items: seals, covers, cables, fasteners, cases, drying materials, consumables and replacement interfaces. The lesson worth borrowing is not romance about expedition grit. It is that remoteness magnifies the operational power of small, failure-prone families of parts.

A second example can be taken from aviation maintenance culture more broadly. Mature aviation organisations learned long ago that repeated write-ups around one part of the system often point beyond that part itself. An intermittent connector issue may implicate moisture control, strain relief, installation practice and environmental exposure. Recurrent seal degradation may implicate cleaning chemistry, storage method and inspection intervals rather than the seal alone. The relevance to Arctic unmanned systems is direct. Good spares forecasting should ask not only what failed last time, but what usually travels with that failure in the real world. Otherwise, the programme stocks replacements for the symptom and not for the neighbourhood in which the symptom lives.

This is where cold-related wear deserves sharper treatment. Cold changes not only performance in use but also behaviour in handling. Materials harden or lose compliance. Small tolerances become harder to read by feel. Parts that rely on subtle seating or positive engagement become easier to misassemble without anyone consciously noticing. Repeated thermal cycling alters the trustworthiness of some interfaces long before outright fracture appears. A spare plan built on temperate wear assumptions will therefore understate the demand for those parts whose principal suffering occurs not during the mission's glamorous phase but during assembly, disassembly, inspection and storage. The Arctic does not distinguish politely between operational wear and maintenance wear. Forecasting should not either.

Contamination is equally important because it creates hidden demand. Fine snow, salt spray, grit, residue from cleaning, condensed moisture and icing by-products all encourage replacement decisions that may not appear in a neat engineering model of service life. The part is not worn out in a textbook sense. It is no longer trusted. That distinction matters. Maintainers in harsh environments often quarantine or replace parts because uncertainty itself has become too expensive to carry into the next sortie. A connector cap with doubtful cleanliness, a seal with ambiguous compression, a port cover that has seen repeated rough handling, a wiping surface that no longer looks protective enough - these items enter the spare flow not because they are physically destroyed, but because the next mission deserves more certainty than they can now honestly offer. A forecasting system that accounts only for obvious breakage will always be late.

This is why forecasting should follow the mission cycle rather than the parts manual. The parts manual tells you what exists. The mission cycle tells you what gets stressed, dirtied, chilled, hurried, reseated, packed, dropped, cleaned and doubted. Those are different maps. Show a realistic Arctic operating rhythm and the contrast becomes obvious. Launch and recovery create repeated handling on exposed connectors and covers. Battery swaps create wear on contacts and housings. Sensor cleaning pressures windows, gaskets and mounting interfaces. Transport and repacking stress latches, padding, cable strain relief and casing integrity. Shelter scarcity may mean some tasks are done in less-than-ideal conditions, which increases the replacement rate of parts that would have been preserved better in controlled environments. Forecasting that follows the manual sees items. Forecasting that follows the mission sees families, touchpoints and attrition loops.

Still, a fair limitation must be stated. Not every cluster deserves equal stocking weight, and not every programme can carry every conceivable family of related spares into the field. Weight, volume, cost and shelf-life all matter. Some parts will fail rarely enough that forward positioning them is wasteful. Others may be so delicate or configuration-sensitive that the better decision is to centralise replacement rather than multiply field stock. It would be simplistic to suggest that the answer is merely to carry more of everything.

But that limitation strengthens the main point rather than weakening it. Because spares are always constrained, the quality of forecasting matters even more. The question becomes which families of failure are most likely to create operational choke points if not stocked. That is a different and better question than "Which individual parts fail most often?" A part that fails modestly but strands the aircraft when absent deserves attention. A family of seemingly minor items that degrade together under cold handling may deserve more forward stock than a larger component whose failure is rarer, more visible or more easily deferred. Harsh-environment forecasting is therefore not a contest in counting. It is a discipline of consequence.

This has implications for how field data is read. Programmes often log failures at the part level because databases like discrete events. Real maintenance experience in severe environments is messier. It includes ambiguous feel during inspection, repeated reseating, suspicious contamination, shortened confidence intervals, families of "probably fine" items that keep becoming "replace to be safe", and clusters of write-ups that only reveal their pattern when viewed across sorties and seasons. A mature sustainment system learns to see these as signals of grouped attrition. An immature one waits until the same family produces a cleaner, more embarrassing failure. Forecasting improves when reporting language is rich enough to capture neighbourhoods of weakness rather than only dead parts.

The quiet motif here is the drawer. Every good maintenance space has a drawer full of small things that outsiders barely notice until one of them is missing. O-rings, caps, clips, seals, short leads, retention hardware, covers, wipes, sacrificial interfaces, cleaning materials. The drawer looks humble. It often decides whether the aircraft flies truthfully. In the Arctic, the drawer is not a convenience. It is the physical expression of a programme's realism about how machines actually decay between bold claims and cold fingers.

This realism also protects against a common organisational error: treating repeated small-part demand as evidence of nuisance rather than design information. If connectors, seals or related items are being consumed quickly under real field conditions, the right response is not only to buy more. It is to ask what the pattern says about maintainability, exposure, packaging, glove compatibility, contamination pathways and the honesty of the original operating concept. Spares forecasting at its best does not merely feed the machine. It feeds back into design. A family that fails together should be stuck together, but it should also be understood together.

The deeper point is that spares planning is a theory of failure written in inventory form. It tells you what the organisation believes will go wrong, in what shape, and with what operational consequence. In temperate, well-supported environments, errors in that theory can sometimes be corrected quickly and quietly. In the Arctic they become visible much faster, because resupply is slower, repair windows are thinner and the cost of carrying doubt across a sortie boundary is higher. The wrong spare forecast is therefore not just a logistical inconvenience. It is a false account of how the system really lives and wears in cold reality.

The unresolved question, then, is not whether a programme has enough parts in some general sense. The sharper question is whether it has learned to stock for the way failure actually clusters under cold, contamination and repeated handling - or whether it is still counting components one by one while the environment keeps teaching that connectors, seals and their quiet relatives rarely fail alone.

Tools, shelters, and repair realism

In November 1915, after the *Endurance* was crushed in the Weddell Sea, Ernest Shackleton's expedition learned a lesson that still applies to autonomous systems in polar operations: improvisation can preserve life, morale and some measure of mobility, but it cannot magically recreate the workshop conditions on which delicate machinery depends. The men salvaged, adapted, stitched, cut and repurposed. They did not repair the expedition back into its original form. The ice had already changed what counted as possible.

That is the right place to begin thinking about field repair in the Arctic. People are naturally drawn to stories of ingenuity under pressure, because ingenuity is vivid and admirable. But austere maintenance is not a morality play about grit defeating circumstance. It is an argument about thresholds. Some interventions remain plausible in the field. Others become progressively less honest as temperature drops, moisture rises, tools stiffen, shelters shrink and the machine's tolerances ask for steadier hands and cleaner surfaces than the environment will permit. Cold changes what hands know.

This is why repair realism matters more than repair optimism. Show the ordinary scene and the distinction becomes severe. A vehicle returns with contamination around a sensor mount, a suspect connector interface, an enclosure that may have taken moisture, and hardware that now needs more than a simple swap to restore trustworthy service. In a proper bench environment, these tasks might be manageable. In a field shelter with gloved hands, poor drying, intermittent heat, cramped work surfaces and the possibility of wind-driven contamination each time the case opens, the same tasks change character. Name the issue clearly and the engineering question sharpens: not whether a fix is theoretically possible, but whether it can be performed repeatably enough to deserve operational confidence. Explain it, and the implication follows. In the Arctic, the difference between repairable and maintainable can be the difference between a sortie recovered honestly and a failure deferred into the next launch.

That is why certain repairs should be designed out rather than trained around. Training has obvious value. Teams must know how to inspect, swap modules, isolate faults, clean surfaces, manage power and preserve evidence of what went wrong. But there is a recurrent institutional temptation to use training as a substitute for maintainable design. A difficult connector reseat becomes a technician skill problem. A delicate seal replacement becomes a drill. A precise calibration sequence in poor conditions becomes a matter of discipline. This is often a category error. When a procedure relies on fine tactile judgement, contamination-free surfaces, adhesive cure behaviour, subtle torque feel, or alignment that cannot be verified confidently in the field, then repetition in training does not abolish the underlying problem. It merely rehearses the circumstances under which a person may be blamed for a task the design should not have required outside a controlled environment.

This is where tools become more than accessories. A tool is a promise about what the environment will let a human do. Precision drivers, torque tools, optical cleaning kits, multimeters, battery warmers, portable lighting, antistatic precautions, drying setups and sealed work mats all embody assumptions about space, heat, dexterity and time. In temperate and well-supported operations those assumptions are often invisible. In the Arctic they become brutally visible. A torque value that is easy to feel with warm bare hands becomes less trustworthy through heavy gloves or numb fingers. A tiny fastener that would be a trivial nuisance indoors becomes a sortie-ending loss if dropped into snow or onto a wet deck. A cleaning kit that works well in a clean workshop can become a source of fresh contamination if used in a cramped shelter where moisture is still moving through the air. The tool does not merely assist the repair. It defines what sort of repair remains truthful.

Shelter matters for the same reason, and perhaps more. It is common to think of shelter as a convenience that makes work faster or more comfortable. In harsh environments it is closer to a machine for preserving the meaning of maintenance. A heated, lit, dry and organised shelter allows certain tasks to remain real tasks rather than hopeful gestures. It preserves the possibility of controlled inspection, careful handling, disciplined drying and the sort of cognitive calm in which people can still tell the difference between a clean fix and a hurried one. Remove that shelter, or reduce it to a nominal windbreak and a heater fighting condensation, and the maintenance universe contracts sharply. The same team, the same tools and the same aircraft may now be limited to module exchange, basic cleaning and conservative fault isolation. The environment has not merely slowed the work. It has changed which repairs can still claim integrity.

A useful historical and contemporary contrast appears in the MOSAiC expedition aboard the *Polarstern* in 2019 and 2020. The expedition's central Arctic location was as unforgiving as any operator could wish to study, yet the vessel's heated internal spaces, workshops and organised technical infrastructure allowed work that would have been unrealistic on sea ice or at a bare forward site. That contrast is instructive. It does not disprove the severity of Arctic repair. It proves that repair realism depends on what sort of environment the organisation has actually built around the machine. When controlled shelter exists, field maintenance expands. When it does not, delicate repair tasks collapse quickly into wishful thinking. The issue is not courage. It is the physics of trust.

A second example comes from aviation and remote scientific operations more broadly, where field teams in Antarctica and the High North have long distinguished between line maintenance and the sort of deeper intervention that must wait for a proper facility. Publicly described polar logistics repeatedly reveal the same lesson: harsh environments do not reward organisations for pretending that every fix can be pushed forward. They reward the ability to classify honestly what can be done where. Swapping a known-good module, cleaning a sensor face under controlled conditions, reconditioning a battery, replacing a robust access cover or verifying a connector family may be sensible. Reworking adhesive bonds, fine soldering, detailed seal rebuilds, precision optical alignment or intrusive electronics repair in a damp, thermally unstable shelter may not be. Maturity begins when those boundaries are accepted before the mission, not argued over after the aircraft is already due back in the air.

This is also why repair realism belongs inside acquisition and design, not only sustainment doctrine. A system that demands frequent fine adjustment in austere conditions is not simply high maintenance. It is badly matched to the theatre. An airframe that forces maintainers to break critical seals for routine access, a payload whose alignment can shift under ordinary handling but cannot be verified easily in the field, or an enclosure that invites condensation around components requiring delicate cleaning is not a neutral engineering choice. It is a decision to spend human skill and shelter quality every time the machine is touched. The Arctic makes such spending expensive very quickly. Better design often looks dull: fewer intrusive procedures, more replaceable subassemblies, clearer condition indicators, less dependence on adhesives or fragile interfaces outside controlled repair, and fewer tasks that ask a tired maintainer to perform jewellery work in survival conditions.

A fair limitation should be admitted. Not every delicate procedure can be designed out, and not every austere site is equally primitive. Some organisations will deliberately invest in mobile shelters, modular workstations, heated cases and disciplined tool chains precisely because their missions demand a wider band of field maintenance. In such cases, repairs that would be reckless on a snow berm may be entirely reasonable on a well-equipped vessel, in a heated container workshop or at a permanent Arctic facility. It would be simplistic to say that delicate work is never supportable outside a factory or home base.

But that limitation only sharpens the main argument. The question is not whether field repair can be made more capable. It is whether the system's maintainability claims are matched honestly to the shelter, tools and human conditions that will really be present. If a repair requires a level of dryness, cleanliness, dexterity and verification that only a minority of sites can provide, then it should not be treated as a generic field task. If a mission concept depends on frequent completion of such tasks at austere points, the concept is carrying hidden fragility. The reconciliation is therefore straightforward. Expand field capability where it is worth the logistics and infrastructure burden. Design away the rest. What should never happen is to pretend that training alone can bridge the gap between the procedure in the manual and the procedure in the wind.

This has consequences for spares planning as well. A part that can only be installed correctly in a proper shelter is not simply a spare. It is a base-level repair demand. A component that nominally supports field replacement but requires post-fit calibration beyond what the site can verify is not truly field-replaceable in the operational sense. The same is true of consumables. Cleaning materials, protective covers, drying aids, work mats, light sources and fastening aids are not peripheral. They are part of the repair envelope. An organisation that stocks the component but not the conditions of its truthful installation is not maintaining a system. It is curating the appearance of preparedness.

The quiet motif here is the workbench. A real workbench is not just a flat surface. It is accumulated order: light, reach, stability, tool discipline, clean space and the assumption that the object under repair will remain still long enough for careful judgement to matter. Much of Arctic maintenance consists of trying to recreate enough of that bench in places determined to deny it. The more a repair depends on the full bench and the less the field can honestly provide it, the more that repair should be treated as an indictment of design ambition rather than as a test of technician character.

This is why maintainability under austere conditions should be judged by sterner criteria than "field serviceable" or "modular". A serious reader should ask whether a repair can be performed with gloved or cold-affected hands, with the available light and shelter, without contaminating adjacent systems, without requiring adhesive or cure behaviour the environment disrupts, without demanding fine calibration that cannot later be verified, and without consuming so much time and concentration that the maintenance queue itself becomes the next failure mode. These are not matters of convenience. They define whether the machine belongs to the theatre or merely visits it on borrowed confidence.

The unresolved tension remains where organisations most like to flatter themselves: at the point where the team is competent, the tools are respectable, the shelter is just adequate enough to tempt persistence, and the aircraft is one delicate intervention away from returning to service. At that moment, is the programme still performing truthful maintenance, or has it crossed into the more dangerous territory where admirable field effort is being used to defend a repair that should never have been required there in the first place?

Corrosion, moisture, and storage discipline

In 2000, when the Russian submarine *Kursk* was recovered from the Barents Sea, investigators were confronting not only the violence of the original loss but the quieter authority of salt water, trapped moisture and time. Metal, wiring, seals and compartments had continued changing after the decisive event was over. The machinery had not been waiting politely for human attention. It had remained inside a process. Decay keeps working.

That is the right place to begin any serious account of corrosion, moisture and storage in Arctic autonomous systems. Operators naturally think in terms of flight time, mission time and maintenance time, as though the machine alternates between action and pause. The Arctic does not respect that distinction. Salt spray, condensation and freeze-thaw cycling continue altering the aircraft when it is on the bench, in its case, on a deck, in a shelter or in transport. A platform can be technically idle and physically busy. What looks like downtime may be the most active part of the failure story.

This is why "non-operational" time is such a misleading phrase. Show the ordinary sequence and the problem becomes concrete. A vehicle recovers from a maritime or littoral sortie with fine salt residue on its surfaces and dampness around access points. It is brought into warmer air. Moisture condenses where the crew cannot easily see it, perhaps around connectors, inside housings, beneath protective covers, along cable entries or on cold-soaked electronics that have suddenly crossed into a more humid environment. It is wiped down, packed, perhaps judged good enough for later attention. Then it cools again, and whatever moisture or salt remained begins its quieter work. Name the governing issue and the argument sharpens at once: storage is an environmental phase, not an administrative interval. Explain it, and the implication becomes severe. A sortie that ends cleanly may still have started the next failure.

Salt spray deserves special respect because it is both obvious and deceitful. Everyone understands that salt and metal are a dangerous pairing. What is less often appreciated is how little salt it takes, and how many paths it has into the life of a machine. A maritime Arctic platform need not be drenched to become vulnerable. Aerosolised salt, deck wash, glove transfer, contaminated cases, damp straps and residue on recovery equipment can all move salt into places where it is not immediately dramatic enough to demand overhaul. It sits in seams, on fasteners, around contacts, under protective lips, in cable strain reliefs, and on the outer edges of seals whose next compression cycle will drive it further inward. Corrosion is rarely theatrical at first. It begins as permission.

That permission matters especially for electronics, because corrosion does not need to destroy a circuit board outright to become operationally expensive. Show a typical field symptom and the pattern is familiar. A connector becomes intermittent only in certain temperatures. A sensor boots normally but behaves erratically after exposure. A control surface or gimbal shows occasional hesitation that cannot be reproduced in the workshop. A battery communication line or health monitor reads strangely only after certain storage histories. None of these symptoms necessarily points cleanly to corrosion in the first instance. That is exactly the problem. Moisture and salt often create faults that masquerade as software unreliability, vibration, poor assembly or operator error. The chemistry travels disguised as diagnosis.

Condensation is often even more treacherous because it feels like a side effect rather than a primary hazard. Yet Arctic operations are full of transitions across temperature and humidity boundaries: cold aircraft into warmer shelters, warm batteries into cold air; cases closed in one environment and opened in another, sensors powered in air that then cools, electronics sealed when dry and later opened when damp. Each transition is an invitation for water to appear where no rain or spray was visibly present. The crew may believe the aircraft stayed dry because nothing dramatic touched it. The platform itself tells a more complex story. Internal surfaces do not care whether moisture arrived through weather, handling or simple thermodynamics. They care only that it arrived.

A useful historical example lies in naval and aviation practice more broadly, where salt, moisture and storage discipline have long been treated as operational rather than housekeeping concerns. Organisations that work routinely from ships or in maritime air learned this lesson the hard way: the environment continues attacking equipment between missions, and preservation routines are therefore as much part of sortie generation as fuelling or route planning. The relevance to Arctic unmanned systems is direct. Littoral cold combines maritime contamination with thermal cycling in ways that make neglect look affordable for a short time and expensive over a season.

A second example can be drawn from polar field science and remote instrumentation, where publicly described programmes often discover that electronics failure is as much a problem of packing, drying and storage as of the original deployment. Instruments returned intact from the field may later exhibit faults because moisture was trapped during recovery or because warming and repacking happened too quickly or in the wrong order. The lesson is not that equipment is fragile in some moral sense. It is that preservation is a sequence, and each step either interrupts or accelerates the chemistry of failure.

Freeze-thaw cycling gives this sequence its own harsh rhythm. Water that enters a small cavity, interface or porous material does not remain inert while temperatures move around it. Freezing can expand, stress or separate. Thawing can redistribute, wet new surfaces or activate corrosion pathways that were paused but not eliminated. Repeated cycles worsen what a single exposure might not yet reveal. This is why the phrase "it dried out later" should make a serious maintainer uneasy. Drying does not erase what freezing, thawing and repeated moisture presence may already have done to coatings, adhesive interfaces, seal compression, plating, or fine conductive surfaces. The damage process is cumulative and often silent until the right combination of temperature, load and time makes it visible.

That silence is what makes storage discipline so consequential. Storage is not simply finding somewhere to put the aircraft until the next mission. It is a set of decisions about whether salt remains active, whether moisture is trapped or released, whether components are allowed to equalise slowly or are pushed through rapid thermal transitions, whether packing materials protect or retain dampness, whether cases become protective shells or chemical incubators. A poorly chosen case can preserve contamination beautifully. A rushed packing routine can seal uncertainty next to electronics more effectively than any hostile environment ever could. The machine does not know it has been "stored". It only knows the conditions it has been left in.

This is also why packing practices deserve more seriousness than they usually receive. A platform returned from cold operations often passes through human hands at the exact point when fatigue, weather pressure and schedule pressure are all encouraging shortcuts. Covers go on quickly. Cases are shut while the machine is still thermally uneven. Damp cloths, absorbent materials or contaminated protective items may share space with sensitive equipment. Battery compartments, ports, windows and access panels may be closed before their local condition has actually stabilised. Each of these small acts has a future. The next sortie may depend on how honestly the previous recovery was packed away.

Still, a fair limitation must be admitted. Not every Arctic operation is maritime, not every storage cycle is equally damaging, and not every trace of moisture or salt becomes a system-threatening event. Some platforms are enclosed well enough, coated well enough and handled carefully enough that storage discipline can be relatively straightforward. Good materials and thoughtful sealing matter. Controlled shelters help enormously. It would be lazy to suggest that every northern mission places electronics in constant silent jeopardy.

But that limitation only clarifies the real standard. The issue is not whether corrosion and moisture are inevitable in every case. It is whether the programme behaves as though they are always possible and operationally significant until proven otherwise. A mature organisation does not wait for visible corrosion bloom or repeatable electronics faults before taking storage seriously. It assumes that salt residue, condensation and thermal cycling are already part of performance, and it structures recovery, drying, packing and inspection accordingly. The reconciliation is simple. Good design reduces vulnerability. Good discipline prevents the environment from cashing in the remaining vulnerability between flights.

This is why corrosion should not be thought of only as a materials problem. It is also a governance problem. Who decides whether the aircraft is dry enough to pack? Who distinguishes cosmetic residue from meaningful contamination? Who owns the transition between recovery and storage? Which signs trigger deeper inspection rather than hopeful release? What logs capture the difference between a dry, clean return and a cold, damp one that merely *looked* intact? These questions matter because corrosion and moisture operate across the seams of responsibility. If storage is treated as nobody's real mission phase, it will become everybody's next unexplained fault.

The quiet motif here is the case latch. It makes a satisfying sound when it closes, a small promise of order restored. Yet that sound can be deceptive. A latched case may protect an aircraft, or it may seal in the ingredients of the next defect. What matters is what was enclosed with the machine: dry surfaces or damp ones, clean interfaces or salted ones, stabilised temperatures or a chemically active transition still underway. The latch ends the visible work. It does not end the process.

A reader trying to judge maintainability under austere conditions should therefore ask harder questions than whether the platform resists corrosion in the abstract. Does the system make it easy to inspect likely moisture paths? Are storage and transport procedures built around thermal transitions rather than only physical protection? Do cases, covers and packing materials help drying and contamination control, or merely improve neatness? Does the organisation treat recovery-to-storage as a managed technical phase, or as a hurried period between the "real" activities of flying and repairing? These questions expose whether the programme understands that preservation is part of performance.

The deeper point is that system performance cannot be confined to moments when the propellers are turning or the payload is collecting. A platform's true operational life includes the bench, the case, the transit rack, the shelter floor and the interval between one decision and the next. Corrosion and moisture exploit every lazy distinction between those phases. They do not care whether humans classify the aircraft as active or inactive. They care whether the environment still has access and time. In the Arctic, it usually has both.

The unresolved question, then, is not whether the machine can survive its missions. Many can. The sharper question is whether it can survive its recoveries, its storage cycles and its pack-outs with the same honesty - or whether the organisation is still treating the closed case and the parked aircraft as the end of operations, when in fact the failure process has already decided to keep working in the dark.

Learning systems and sustainment governance

In January 1986, the Rogers Commission began examining the loss of the Space Shuttle *Challenger*, and one of the most enduring lessons was not only about hardware but about institutions. The catastrophe did not emerge from a world in which signals were absent. Warnings existed. Anomalies had histories. Decisions had pedigrees. What failed, in part, was the organisation's ability to convert scattered technical experience into governed learning before the pattern became lethal. The record had been speaking.

That lesson belongs directly to Arctic sustainment. Remote systems in severe cold do not merely degrade; they narrate their degradation through maintenance notes, repeated workarounds, recurring contamination, shortened battery confidence, suspect connectors, moisture traces, and the quiet migration of "probably fine" into "we keep seeing this". A programme that cannot hear those murmurs across seasons does not merely suffer inefficiency. It becomes institutionally deaf. In the Arctic, where the same environment returns with new combinations of old insults, deafness is expensive.

This is why learning systems should be treated as part of the machinery rather than as an administrative layer wrapped around it. Show the ordinary field rhythm and the point becomes clear. A platform returns with a mildly fouled sensor, a battery that needed longer conditioning than expected, a connector that felt uncertain under glove-handling, and a shelter-side repair that took more judgement than the manual implied. Each event can be logged, dismissed, or remembered only in the mind of the person who performed it. Name the serious option and the argument sharpens immediately: defect information as operational capital. Explain it and the implication follows. If such observations are not gathered, compared, classified and acted upon, the organisation will relearn the same Arctic lesson under different weather and different names.

Maintenance data matters because it is often the first place the machine admits it is becoming a different machine from the one imagined by procurement language or temperate testing. Flights tend to produce clean stories: launched, completed, recovered, aborted, diverted. Maintenance produces messier truths. This connector family is ageing badly. That access panel traps moisture more than expected. This battery batch no longer behaves like its paperwork. These seals survive nominal use but not the actual handling sequence in the cold. These sensors return with residue after a particular recovery geometry. Such knowledge is rarely dramatic at first. Its power lies in accumulation. A learning system gives accumulation a home before it becomes a pattern everyone "just knows" and no one has formally honoured.

This is why defect reporting must be more than a bureaucratic sink for faults. A weak reporting culture produces two equally bad outcomes. In one, crews report only clear failures, omitting ambiguous signs because they seem too minor, too subjective or too familiar to matter. In the other, reports become so generic and defensive that they cannot teach anything beyond the fact that someone was uneasy. Good sustainment governance sits between those extremes. It encourages specificity without turning every anomaly into melodrama. It makes room for phrases like intermittent, condition-dependent, repeat under thaw, appears after glove-handling, not yet failed but confidence reduced. Those are not soft observations. In severe environments, they are often the true early-warning system.

A useful historical example lies in civil aviation's development of safety reporting and maintenance record culture, where organisations gradually learned that recurring minor anomalies could carry more safety meaning than isolated dramatic failures. The point was never merely to collect more data. It was to connect reports, maintenance actions, environmental conditions and configuration history tightly enough that weak signals became usable before they matured into accidents. The relevance to Arctic unmanned systems is plain. The theatre is different. The institutional challenge is not. A repeated winter connector issue recorded as three separate inconveniences is poor governance. The same issue recognised as a cold-handling failure cluster becomes design information.

A second example can be drawn from polar field science and remote technical operations more broadly, where seasonal return is both a gift and a trap. It is a gift because the environment repeats enough to allow disciplined comparison. It is a trap because organisations often treat each season as a fresh campaign rather than as the continuation of an unfinished argument with the same weather, moisture, salt, light and handling realities. Publicly described remote programmes tend to mature when they stop treating last year's defects as mere archive and start treating them as present-tense constraints on configuration, stocking, training and release decisions. The Arctic remembers. Good governance behaves as though it does.

Configuration control is the hinge that lets maintenance data become organisational learning rather than anecdote. Without it, teams can notice everything and still fail to improve. If one aircraft carries a revised seal material, another a different connector retention feature, a third a software patch altering battery conditioning logic, and a fourth an unofficial field modification applied under pressure, then comparing outcomes becomes murky very quickly. Configuration discipline is what prevents a programme from asking, season after season, whether the environment changed when in fact the fleet changed without enough traceability. It sounds dry because it is dry. It is also one of the few ways to stop the Arctic from turning local improvisations into long-range confusion.

This is why institutional maturity shows itself so plainly in the relationship between field fixes and formal configuration. Immature programmes tolerate a great deal of tacit adaptation. A technician learns a better packing method. A crew starts carrying unofficial absorbent material in a case. Someone slightly alters a handling sequence for a stubborn connector. A shelter team adds a local safeguard to battery warming. These workarounds may be intelligent and necessary. They become dangerous when they remain socially transmitted but organisationally invisible. Mature governance does not crush such initiative. It captures it, tests it, formalises it where justified, or retires it where it solved one problem by quietly creating another. That is how a programme graduates from folklore to engineering.

Still, a fair limitation deserves emphasis. Data does not teach automatically. Organisations can collect mountains of maintenance information and remain poor learners if the data is fragmented, misclassified, delayed, stripped of environmental context or trapped in systems no one uses to make decisions. Worse, badly designed governance can punish reporting by associating every anomaly with blame or administrative burden. In that atmosphere, people quickly learn to understate uncertainty, simplify narratives and keep the most operationally useful knowledge local and unofficial. A learning system can therefore fail by excess as well as by neglect. The problem is not too little paperwork or too much. It is whether observation is actually being turned into changed behaviour.

But that limitation only clarifies the standard. Good sustainment governance links three things tightly: observation, decision and memory. Observation means maintainers and operators can record what the environment is doing to the machine in language rich enough to preserve significance. Decision means those records shape stocking, intervals, shelters, release standards, training and design revision rather than merely populating a database. Memory means the organisation can compare winters to winters, platforms to platforms, and configuration states to outcomes without relying on who happens to still be on the team. This triad is what allows improvement across seasons. Without it, every deployment begins with an amnesiac optimism the Arctic is well practised at punishing.

The safety culture dimension matters because sustainment governance is one of the most revealing tests of what an organisation truly values. Almost any programme can claim to care about reliability. Fewer are willing to ground an aircraft because maintenance evidence has become suggestive without yet being cinematic. Fewer still are willing to admit that a recurring small defect points not to technician weakness but to a mismatch between system design and actual field conditions. A strong culture does not romanticise maintenance as heroic improvisation. It treats it as evidence production. That changes the moral status of the maintainer. The person at the bench is not merely returning hardware to service. That person is helping the organisation decide what it really knows.

The quiet motif here is the ledger. A serious sustainment organisation keeps one, whether literally or through disciplined digital systems, because the ledger is where the machine's seasons accumulate into institutional memory. Each defect note, each component change, each shortened interval, each repeated contamination event, each unofficial workaround either enters that ledger or is left to evaporate in conversation. The difference is profound. What stays only in speech cannot govern next year's release decision with the same authority as what has been made traceable, comparable and contestable. The ledger is not romance. It is how the cold is translated into policy.

This has direct consequences for availability as well. A fleet that learns can reduce unnecessary conservatism where evidence supports it and tighten discipline where evidence demands it. A fleet that does not learn oscillates between the two worst habits: optimism that ignores repeated warning signs, and blanket caution that treats all uncertainty as equally paralysing because nothing has been analysed sharply enough to distinguish true threat from manageable nuisance. Both habits waste aircraft. Both waste trust. Learning systems are what let a programme become more precise across seasons rather than merely more nervous or more complacent.

The unresolved test comes when a familiar anomaly appears again, small enough to wave away and common enough to feel routine. A connector seats with less confidence than it should. A battery takes longer to condition. A sensor needs more cleaning than last season's baseline would once have tolerated. At that moment, does the organisation possess a sustainment system capable of turning repetition into redesign, stocking change, interval revision or procedural correction - or is it still treating maintenance as local housekeeping, while the Arctic patiently composes the same lesson for another year?

Chapter 10

Launch and Recovery in High Winds

What if the most honest way to judge an Arctic drone is not by how far it can cruise or how long it can remain aloft, but by whether it can come back under conditions that strip away the fiction of easy control? That question cuts through a great deal of inflated language. Cruise performance flatters. Recovery humiliates. A platform can look elegant in transit, efficient in surveillance and technically impressive in its range, yet still remain operationally dubious if the final moments of the mission depend on calmer air, wider margins or steadier surfaces than the theatre is willing to provide.

That is why launch and recovery deserve a harder intellectual status than they usually receive. People tend to treat them as bookends to the real mission, a brief technical preface and a practical ending. In high winds, they are closer to an audit. Every assumption made elsewhere in the system comes due at the edge of the deck, at the top of the catapult rail, in the flare to a strip, in the hover above a moving point, in the split second when control law, sensor lag, gust structure and human judgement must stop sounding persuasive and start being enough. The runway tells the truth.

Severe wind makes this especially unforgiving because it turns local geography into behaviour. Air does not arrive as a clean, constant adversary. It bends around ships, ridgelines and ice features. It accelerates, separates, rolls and returns. The machine may remain nominally inside broad environmental limits while still entering a pocket of air that its tuning, sensing or recovery logic interprets too slowly or too simplistically. A concept that looks viable in calm or moderately variable conditions can therefore prove fragile at the exact point where the mission must either resolve safely or admit that it should never have been attempted in that form. The hardest weather is often the weather nearest the ground.

The central tension lies there. Operators want a launch and recovery method that minimises footprint, reduces support burden and preserves tactical or scientific flexibility. The environment demands something rougher: margin, tolerance, repeatability and a way to abandon the attempt without turning prudence into loss. Those demands do not always align. A method that looks efficient on paper may import hidden costs in sensor precision, deck preparation, approach geometry or diversion planning. A method that appears logistically awkward may be the only one honest enough about wind. High-wind operations therefore force a choice not between sophistication and simplicity, but between performance claims and recoverable truth.

That choice matters far beyond a single sortie. Recovery design shapes fleet credibility, operator behaviour and institutional candour. If crews know that a platform's return phase is narrow, fussy or overconfident, they will begin quietly altering routes, shortening tasks, preserving extra energy and carrying private doubt that no specification sheet records. If managers refuse to admit that recovery is the real limit, they will misread losses and wave-offs as isolated setbacks rather than as evidence that the whole concept has been balanced around the wrong moment. Mission availability, in this sense, depends not only on aircraft survivability but on whether the organisation has priced the tax recovery imposes on every launch from the outset.

The stakes also become moral once aborts enter the picture. A system asked to persist into a marginal recovery is not merely taking aerodynamic risk. It is risking confused accountability. The language of autonomous recovery can hide this by making the final phase sound like an extension of cruise logic rather than a crisis of shrinking options. Yet an abort in high wind is never only a technical branch. It raises questions of authority, diversion, remaining energy, alternate surfaces, communication quality and what kind of evidence the aircraft can preserve if it cannot be brought back cleanly. A platform that cannot retreat safely is not operationally mature, however elegant its return algorithm may appear in demonstration footage.

The path ahead follows this compression of truth. It begins with the physical forms launch and recovery can take, and with the uncomfortable fact that recovery method often decides concept viability long before cruise performance enters the conversation. It then moves into the disturbed air around ships, ridgelines and ice edges, where local flow, sensor delay and control margins reveal what the machine really knows how to handle. From there the argument tightens around aborts and diversion, where survivability and accountability begin to matter more than elegant completion, and where the difference between a recoverable wave-off and a disguised loss becomes impossible to ignore.

A useful way to hold these questions together is to think of a doorway in strong wind. Walking through it is one thing. Carrying something fragile through it is another. The doorway concentrates forces, changes footing and punishes clumsy timing. Launch and recovery do the same to autonomous systems. They compress the whole mission into a narrow frame where every small deficiency in sensing, tuning, preparation, authority design and environmental honesty suddenly weighs much more than it did out in open air.

The question that remains is not whether the aircraft can be made to launch or recover under some favourable interpretation of high wind. Many can. The sharper question is whether the chosen method, the disturbed air it must survive, and the abort logic that sits behind it all describe a system that can return without bluffing - or whether every mission has quietly been counting on a doorway that only looks wide enough until the gusts begin.

Runway, catapult, and vertical options

In April 2021, NASA's *Ingenuity* rose from Jezero Crater and proved something that engineers had long understood in principle but the wider public often forgets: how a vehicle leaves the ground and how it returns are not mirror images of the same problem. Vertical flight removed the need for a runway altogether, but it did not abolish recovery discipline; it merely relocated it into a far tighter relationship between control authority, surface conditions and energy margin. The machine flew. The bill remained.

That is the right place to begin any serious comparison of runway, catapult and vertical options in harsh weather. Launch methods are often discussed as though they express platform character - sleek runway aircraft, efficient catapult systems, flexible vertical vehicles - while recovery is treated as a practical epilogue. In the Arctic, that hierarchy collapses. Recovery is not the ending after the interesting part. It is the test that decides whether the concept was honest all along. A platform may cruise beautifully and still be operationally compromised if its method of returning demands calmer air, cleaner geometry or steadier surfaces than the theatre routinely provides.

This is why every launch method must be judged by a combined question: what footprint does it require, what risks does it concentrate, and what weather does it ask the organisation to normalise? Show the practical problem and the distinctions become immediate. A runway aircraft may deliver admirable efficiency in cruise and relative simplicity in some phases of recovery, but it asks for prepared surface, approach discipline and enough lateral and longitudinal space for wind to remain a manageable variable rather than a sudden moral argument. A catapult-launched aircraft may solve the launch footprint elegantly and free the programme from dependence on a full strip, yet it still has to come back somehow, and the chosen recovery method may import its own severe assumptions. A vertical aircraft looks liberating because it can rise and settle without large ground infrastructure, but in high winds that apparent freedom often becomes a demand for exquisite control in the most disturbed air closest to the surface. The method is never just a method. It is a contract with the local atmosphere.

Runway operations illustrate this clearly because they flatter planners with a kind of familiar realism. Everyone understands a strip. It feels orderly, geometrically honest, anchored to the old grammar of aviation. That familiarity is valuable. It can also seduce. In high winds, the runway is not simply a place to land but a demand for stable enough alignment, manageable drift, believable surface condition and a recovery path that still means something once gusts, crosswind and poor visibility begin to interact. The runway method therefore often looks safer than it is because its infrastructure is visible while its weather tax is paid gradually, in widening margins, more conservative dispatch decisions, reduced payload tolerance and the quiet exclusion of days that had seemed operationally possible from a cruise-performance perspective.

Catapult systems solve a different class of problem and create a different class of temptation. They are attractive precisely because they seem to decouple performance from runway dependence. A vehicle can be pushed briskly into useful airspeed from a smaller footprint, from a ship, from austere ground or from awkward terrain where a proper strip would be a fantasy. In that sense catapults are often operationally serious devices. Yet their seriousness is frequently misread because people judge them by the efficiency of launch rather than by the honesty of recovery. A catapult can make an aircraft look fieldable in places where its return method remains the real burden. Net recovery, skid landing, belly landing, arresting system or alternative site diversion - each brings its own vulnerability to wind, surface, timing and repeatability. A clean launch can therefore disguise a very expensive homecoming. The tax is merely deferred.

Vertical options are the most rhetorically seductive because they seem to promise emancipation from infrastructure altogether. No strip. No rail. No deck run. Just lift, hover, reposition and settle. In benign conditions that promise has real force. In high winds it becomes more conditional than many public discussions admit. Near the ground, vertical systems often inhabit the most aerodynamically confused part of the mission envelope: rotor wash interaction, disturbed flow from structures or terrain, rapid gust change, poor visual reference, and very small margins for control lag or overcorrection. A vertical aircraft is not exempt from the theatre's weather tax. It often pays it in tighter energy budgets, stricter hover authority requirements, more vulnerable sensor assumptions and a much narrower tolerance for local turbulence than its apparent versatility suggests.

A useful historical example lies in shipboard aviation more broadly, where navies learned long ago that launch method and recovery method are not symmetrical virtues. A system may leave a deck with confidence and still prove operationally awkward if the returning phase depends on conditions the ship, sea and air cannot reliably provide. The relevance here is not to borrow glamour from carrier operations. It is to notice the structural lesson. Launch can be engineered into a crisp mechanical act. Recovery remains a continuing negotiation with wind, motion and timing. Any Arctic concept that admires the first and understates the second is building its case on the wrong half of the mission.

A second example appears in the public record of small unmanned systems used from vessels and austere sites, where apparently elegant cruise platforms have often been constrained less by their endurance or sensor value than by the practical honesty of bringing them back. Programmes discover, sooner or later, that a launch method solving one logistics problem can create a recurring recovery tax in another currency: higher crew workload, more frequent wave-offs, tighter weather windows, more difficult maintenance after rough recoveries, or a dependence on alternate sites that complicates accountability and data custody. What matters is not that one method is universally superior. What matters is that the true operational cost sits where the aircraft re-enters the human world.

That is why footprint must be understood in more than geographical terms. A runway has a physical footprint, obviously. So does a catapult, a recovery net, a landing pad or a ship deck. But each method also has a *weather footprint*: the slice of real atmosphere in which it remains repeatable enough to deserve planning confidence. Vertical systems often claim a small physical footprint while carrying a surprisingly demanding weather footprint near touchdown. Runway systems require larger physical preparation but may, in some circumstances, offer a more legible final approach. Catapult-launch concepts often compress ground footprint on departure only to expand recovery footprint in ways that logistics and doctrine must absorb later. A mature programme studies both maps. An immature one admires only the smaller pad or shorter launch lane.

Still, a fair limitation should be admitted. No method is condemned in the abstract. Vertical systems can be remarkably valuable where deck space, terrain or tactical exposure make other options unrealistic. Runway aircraft can be entirely sensible where prepared surfaces and approach geometry are honest and supportable. Catapult arrangements can enable useful operations from places that would otherwise exclude fixed-wing performance altogether. The issue is not choosing a universal winner. It is refusing to let cruise efficiency or launch cleverness obscure the full recovery burden. Some concepts are viable precisely because their organisations have accepted and paid that burden properly.

But that limitation only clarifies the deeper point. Viability is not an intrinsic property of the airframe or launch device alone. It is a relationship among method, weather, footprint, crew discipline, diversion logic and the quality of the final approach the environment is willing to lend. A concept that looks superior by range, payload or transit efficiency may still be the wrong one if its recovery tax is habitually paid through narrow margins, optimistic abort thresholds or quietly accepted damage and attrition. Conversely, a platform that seems less glamorous in cruise may be the more serious operational choice if its return path is broader, duller and more repeatable. In high-wind Arctic work, dull recovery is often a mark of maturity.

The quiet motif here is the receipt. Missions are often priced during planning by distance, endurance, fuel, battery and payload opportunity. The receipt arrives later, at recovery, where the hidden charges appear: deck motion, gust spread, surface uncertainty, operator workload, sensor lag, diversion distance, post-recovery maintenance, and the simple question of whether the aircraft can be brought back in a state that still deserves trust tomorrow. A launch method is therefore never fully paid for at launch. Its real price is settled when the machine comes home - or does not.

This is what makes recovery the decisive tax on every mission. The reader should not ask merely which option gets the aircraft into the air most efficiently. The harder question is which option allows the organisation to remain honest about weather tolerance, footprint and survivability once the atmosphere has become uncooperative and the mission has already spent much of its margin. If a platform can launch from almost anywhere but recover only from optimism, then the concept is not flexible. It is indebted.

The unresolved test comes when a system looks excellent in transit, the launch arrangement seems ingenious, and the weather has begun to tighten only at the edges. At that point, does the chosen method still describe an aircraft that can be recovered with ordinary discipline, or has the entire mission been quietly borrowing against a return phase whose tax the organisation may not really be able to pay?

Deck motion, gusts, and control margins

On 24 October 1963, the aircraft carrier USS *Enterprise* operated in the North Atlantic under conditions that reminded naval aviators of an old and unforgiving truth: the deck is never just a surface. It moves, heaves, yaws under the aircraft, and sits inside air that has already been bent by hull, island structure, sea state and speed through the water. The significance goes well beyond crewed carrier aviation. A moving recovery point turns the last phase of flight into a problem of relative motion and disturbed air rather than simple descent. The air is already crooked.

That lesson matters sharply in Arctic unmanned operations because the worst wind is often not the wind measured in broad area forecasts or even by a masthead sensor. It is the local wind the vehicle actually flies through in the final seconds: accelerated around a ridgeline, separated off a cliff face, curled over a ship's superstructure, or broken by the thermal and geometric complexity of an ice edge. A platform may appear to be operating within nominal conditions and still enter a recovery regime in which the air is changing faster than its sensors, controllers or operators were really designed to interpret. High wind in this sense is not only a matter of strength. It is a matter of shape.

This is why wind shear and turbulence deserve to be treated as recovery conditions rather than as abstract meteorology. Show the practical case and the issue becomes immediate. An aircraft approaches a ship, a small landing pad near a ridge, or a recovery area beside uneven ice. Its position estimate remains plausible. Its descent law is still active. The operator or autonomy stack has a recovery solution. Then the local air changes: one rotor disc or wing section sees a different reality from another, a gust arrives with a new direction rather than simply more force, the surface itself is rising or falling beneath the vehicle, and the platform's sensors are now trying to describe an environment that is no longer changing smoothly enough for their original assumptions. Name the problem and the engineering question sharpens: relative motion in disturbed flow. Explain it and the implication becomes severe. Recovery fails not because the aircraft forgot how to fly, but because the final geometry asked it to solve a faster, noisier problem than it had been authorised to believe existed.

Ships intensify this because they create moving airfields inside moving air. The deck itself is not fixed. It pitches, rolls, heaves and translates. The recovery point therefore has its own motion history even before the wind field around it is considered. Add the vessel's superstructure, which diverts, accelerates and sheds air in patterns that vary with heading and wind direction, and the recovery zone becomes a compact theatre of changing assumptions. A landing system tuned on calm days may perform elegantly when the deck behaves as a nearly steady platform and the flow remains tolerably uniform. The same system, faced with more abrupt gust structure and deck motion, may begin chasing an ever-shifting target. The result is not only harder piloting. It is a shrinking control margin in which each correction arrives into air and geometry already altered from the state that provoked it.

Ridgelines create a related but different cruelty. Air crossing elevated terrain does not politely preserve the nice, broad vectors that planning tools imply. It accelerates over crests, separates on lee sides, tumbles into rotors, and turns narrow corridors into traps for aircraft whose route looked entirely respectable in larger-scale wind logic. In Arctic operations, where launch and recovery sites may be chosen precisely because they offer limited shelter or terrain access, this matters acutely. A site that looks protected at ground level can still sit beneath a hostile transition zone just above it. The aircraft may therefore face a final approach in which the apparent refuge of terrain becomes the producer of the very turbulence and shear that defeat stable descent. The geography that made the site usable begins charging interest at the end.

Ice edges add another form of complexity because they juxtapose surfaces with different thermal and mechanical behaviour. Open water, broken ice, pressure ridges and snow-covered flats do not shape local flow in the same way. The transitions among them can alter turbulence intensity, surface roughness effects and the visual texture on which optical or lidar-based systems depend. This matters more than is often admitted because recovery logic is rarely flying through a purely aerodynamic problem. It is also flying through a sensing problem. A descent controller depends on what the aircraft can measure and infer about height, drift, platform motion and closure rate. If the air becomes more chaotic while the visual or ranging references also become less stable, the platform can lose margin on two fronts at once. Aerodynamics and perception go hand in hand.

This is where sensor lag becomes operationally decisive. A control system does not respond to reality directly. It responds to a representation of reality assembled through sensors, filtering, estimation and command execution. In calm conditions the delay and smoothing involved in that chain may be entirely acceptable. In strongly gusting conditions near a moving or visually ambiguous recovery point, the same delay can become the opening through which oscillation enters. Show the sequence and the mechanism is plain enough. A gust pushes the aircraft. The sensors notice it through noise and filtering. The controller commands a correction. By the time that correction acts, the gust structure has changed or the deck has moved or the visual estimate has shifted. The next correction is then based partly on a state that no longer exists. The aircraft begins to hunt rather than settle. It is reacting accurately to yesterday's air.

That is why a landing system tuned for calm conditions may oscillate when gusts exceed its assumptions, even without any obvious software bug. The tune itself is the hidden assumption. Control gains, filter settings, estimator confidence, surface tracking logic and allowable approach geometry all encode beliefs about how quickly the world changes and how smooth the error should look between one instant and the next. In benign conditions those beliefs are rewarded. In harsh local flow they can turn grace into brittleness. A system that was stable because it corrected gently may become under-reactive. A system tuned for crisp correction may begin overcontrolling because it mistakes transient gust response for persistent drift. The distinction matters because it shows why weather tolerance is not merely a line in a manual. It is built into the very temperament of the control law.

A useful historical example comes from shipboard helicopter operations, which have long taught aviators that deck recovery is governed as much by local air and relative motion as by nominal aircraft performance. The point is not to borrow crewed rotary-wing lore uncritically for unmanned systems. It is to notice how mature that culture has become about the final phase of flight. It does not assume that broad wind limits tell the whole truth. It treats the interaction of ship motion, structure-induced flow, pilot workload and timing as central to recovery viability. The relevance to autonomous systems is direct. Where a human once carried some of that burden through embodied judgement, the unmanned platform must carry it through sensing, control design and abort discipline. The physics did not grow kinder when the pilot left the cockpit.

A second example can be seen in the public record of small shipborne and ridge-adjacent drone operations, where platforms that behave credibly in open-air transit often become erratic in the last moments before touchdown or capture. Operators describe late drift, repeated wave-offs, unstable hover behaviour, or descent cues that seemed trustworthy until the aircraft crossed into the most disturbed zone. The importance of these accounts lies not in any one dramatic mishap. It lies in their pattern. High-wind recovery rarely fails because the aircraft suddenly encounters a generic "bad weather" state. It fails because the final local flow, platform motion and sensing chain together demand more control authority and faster truth than the system actually holds.

Still, a fair limitation should be admitted. Not every ship, ridge or ice edge produces equally hostile conditions, and not every aircraft is equally vulnerable. Good site choice, sensible headings, broader recovery corridors, more conservative altitude management and better sensing can materially improve the situation. Some platforms are also designed with enough excess control authority and enough environmental awareness to tolerate disturbed final air more honestly than others. It would be simplistic to imply that every deck or ridge recovery in wind is structurally deceptive. Sometimes the local flow remains manageable and the system's assumptions remain sound.

But that limitation only clarifies the standard. The issue is not whether recovery in gusty environments is possible. It is whether the machine's control margins remain truthful under the local conditions actually created by ship motion, terrain shape or ice geometry. Truthful here means that the aircraft can sense the disturbance fast enough, estimate it honestly enough and command enough correction without entering a self-created oscillation or a repeated sequence of barely controlled wave-offs. The reconciliation is therefore stern. A platform is not weather-tolerant because it survives broad wind in cruise. It is weather-tolerant only if its final control problem remains within its sensing and actuation integrity near the real recovery point.

This is why control margin should be treated as a living quantity rather than a performance boast. Margin is spent by deck motion, by gust onset, by sensor lag, by visual ambiguity, by low-altitude turbulence, by estimation error and by the simple fact that the aircraft is nearer the point where correction and collision become too intimate. A strong system therefore does not merely carry margin. It knows when that margin is being consumed faster than recovery can justify. That knowledge may force a wave-off, a diversion or a mode change before the aircraft looks visibly out of control. To outsiders, such caution can seem disappointing. In truth it is one of the clearest signs that the recovery concept is operationally honest.

The quiet motif here is the swinging door. A person can walk through a stable doorway without much thought. Put that same doorway on a moving hinge in a crosswind and the act changes completely; timing, force and angle all become matters of consequence. High-wind launch and recovery are like that. The sky is no longer a neutral opening. It is a moving frame through which the aircraft must pass while carrying delayed information and finite authority. Under those conditions, small misjudgements grow teeth quickly.

The unresolved test comes when the broad forecast still looks acceptable, the vehicle remains responsive in transit and the recovery point appears only modestly disturbed. At that moment, does the aircraft still possess enough timely knowledge and enough control margin to settle honestly into the final air, or is it already entering that dangerous regime where the landing law is correcting faithfully for a world that has begun changing faster than the machine can truly believe?

Abort logic and safe diversion planning

On 15 January 2009, US Airways Flight 1549 left LaGuardia and, within minutes, Chesley Sullenberger and Jeffrey Skiles had abandoned the original plan in favour of the only outcome that still honoured the aircraft's shrinking reality: a ditching in the Hudson River, New York. The event belonged to crewed aviation, not Arctic drones, yet its lesson is exact. A mission does not become more serious by clinging to its intended ending after the conditions that made that ending reasonable have disappeared. It becomes more serious by recognising that survivability has replaced elegance as the governing value. The plan had to shrink.

That is the proper starting point for abort logic in high-wind Arctic operations. People often speak of an abort as though it were a pause in success, a regrettable interruption on the way to the real achievement of landing where planned, with the mission intact and the aircraft undamaged. In severe wind, that way of speaking is already dangerous. A wave-off is not a failure of nerve or a software branch of last resort. It is often the clearest sign that the system still knows the difference between a recoverable future and a stubbornly pursued fiction. If the machine, the crew or the organisation cannot make that distinction early enough, then recovery is no longer a phase of operations. It is a gamble in uniform.

This matters because the final moments before landing concentrate all the dishonesty a mission has been quietly accumulating. Wind may now be more disturbed than the broad forecast implied. The recovery point may be moving or visually less legible than planned. Sensor lag that seemed tolerable in transit may now distort the last corrections. Energy that looked generous on paper may be less generous after holds, headwinds or repeated approaches. Communications may still exist while no longer supporting meaningful intervention. Show the scene plainly and the issue becomes harder than mere airmanship: the aircraft is not deciding only whether it can continue descending. It is deciding whether the planned recovery still deserves belief.

Abort logic is the discipline that turns that decision from hope into policy. Name it properly and the argument sharpens: a set of predeclared conditions under which the mission must stop trying to complete itself in its original form. Explain why this matters in the Arctic and the implication becomes severe. Without such logic, every failed approach invites another, every second pass begins to borrow from energy and clarity that are no longer replenishing themselves, and every minute of delay increases the chance that prudence will arrive too late to remain cheap. A wave-off made early can preserve an aircraft, a log, a payload record and an intelligible story. A wave-off made late can still be correct and yet inherit a much poorer set of options. Timing is moral here.

This is why "go around" and "divert" should never be treated as interchangeable ideas. A go-around is a temporary refusal of the present approach. A diversion is the admission that the original recovery concept has ceased to command enough trust. The difference matters because organisations often train the first and romanticise the second without fully preparing for either. Show the usual drift and it is easy to recognise. A platform waves off once, then again, because the deck, strip, pad or surface still appears physically available. Each attempt feels defensible in isolation. Meanwhile, wind remains unstable, local flow remains distorted, and the aircraft's options quietly narrow. What looked like disciplined persistence begins to resemble progressive self-entrapment. Good abort logic breaks that pattern before repetition becomes its own false argument.

A useful historical example lies in the 23 July 1983 diversion of Air Canada Flight 143 to Gimli, Manitoba. The circumstances were utterly different from an Arctic autonomous recovery, yet the structural lesson is relevant. The intended destination stopped being the meaningful centre of the problem. The crew survived by accepting a stranger, less elegant outcome that still preserved life and accountability. The point is not to borrow drama from crewed aviation for its own sake. It is to notice that real professionalism often consists in relinquishing the original endpoint before the price of loyalty to that endpoint becomes catastrophic. Autonomous recovery claims should be judged by the same standard. If the machine can only look competent by continuing towards an increasingly dishonest plan, then its autonomy is performative, not mature.

A second example comes from NASA's Mars helicopter *Ingenuity*, first flown in April 2021 at Jezero Crater. The environment was not windy in the terrestrial sense, and there was no Arctic deck or ridgeline beneath it. Yet the programme's seriousness lay partly in how tightly it bounded action and preserved the right to stop asking the vehicle for more than the evidence justified. That lesson transfers cleanly. Honest autonomy does not prove itself by pressing on whenever the nominal plan remains just barely imaginable. It proves itself by respecting the line at which continued mission completion would consume more trust than the system has earned. In harsh Earth environments, that line often arrives at recovery.

Safe diversion planning begins long before the abort itself. That sentence sounds procedural because it is. A diversion site is not simply another place on the map where the aircraft might technically fit. It is a place whose wind, surface, geometry, access, accountability and post-landing consequences have already been judged against the kind of uncertainty the abort is likely to produce. Show the alternative and the weakness is immediate. A platform leaves an unstable recovery zone and heads for the nearest apparent alternative, only to discover that the site is visually deceptive, operationally inaccessible, or still dependent on the same weak navigation and control assumptions that poisoned the original approach. The aircraft has not escaped the problem. It has moved it sideways. Good diversion planning therefore does not search for another coordinate. It searches for another *truthful future*.

That phrase deserves its weight because diversion is about more than keeping the aircraft out of the water or off the rocks. A serious diversion preserves survivability and accountability together. The aircraft should arrive somewhere that can later be explained, reached, inspected and integrated into the organisation's learning rather than merely counted as "not lost yet". This is especially important in the Arctic, where an intact vehicle can still be operationally expensive if it lands on moving ice, beyond safe reach, or in a place where data custody and configuration knowledge are compromised by the conditions of retrieval. A diversion that preserves the airframe while destroying the clarity of what happened is only a partial success. Autonomous recovery is operationally honest only if it values legible outcomes as well as intact hardware.

Still, a fair limitation must be admitted. Not every wave-off should escalate into diversion, and not every challenging approach proves that the original recovery concept was false. Some systems and crews can manage a second attempt safely when the cause of the first wave-off is understood, bounded and genuinely temporary. Overly rigid abort logic can waste opportunities, drive unnecessary diversions and create its own safety penalties. A machine that gives up too early is not automatically wiser than one that judges carefully and tries again. The issue is not maximal caution. It is whether the reasons for continuing are stronger than the reasons for simplifying, and whether those reasons remain evidence-based rather than emotional or schedule-driven.

That is where control authority and accountability meet. If an aircraft waves off, who decides whether to reattempt, hold or divert? If the system decides, on what basis and with what traceable priorities? If a human decides, through what latency, with what sensor picture and what remaining authority? These are not narrow interface questions. They reveal the programme's underlying ethics. An organisation that prizes mission completion in language but survivability in procedure will eventually be found out by the first genuinely difficult recovery. One that prizes survivability openly will design abort logic, diversion maps, energy thresholds and authority handovers so that the machine never has to feign confidence for the sake of appearances.

This is why "autonomous recovery" should be tested with a hard and slightly unfriendly question: autonomous to do what, and up to which point? A system that can land neatly in prepared, steady conditions may still be deeply immature if its recovery logic cannot recognise when the airfield, deck or pad has become more uncertain than its own control, sensing or authority chain can honestly support. A system that waves off early, preserves options, protects the aircraft's evidence and chooses a dull but reachable diversion may look less heroic. It is often the one behaving with greater integrity. The machine's real character appears when it stops trying to save the plan.

The quiet motif here is the exit sign. In calm conditions, no one thinks much about it. Under smoke, confusion or crowding, it becomes the difference between organised movement and panic dressed up as decisiveness. Abort logic plays the same role in high-wind recovery. It is not there to flatter the mission when all is going well. It is there to remain clear when the air, the sensors and the schedule begin pushing people and machines towards one attempt too many. A visible exit is a form of discipline.

The reader should therefore judge recovery systems not by how confidently they speak of landing, but by how honestly they know when not to. Does the aircraft protect enough margin to make a wave-off meaningful rather than ceremonial? Does it carry preplanned diversion options that preserve not just survival but reachability and explanation? Does its authority design prevent a late, proud continuation from masquerading as competence? Does the organisation treat an uneventful diversion as a professional outcome rather than as a blemish to be quietly rationalised away?

The unresolved question arrives in the final tightening of every hard mission: the aircraft is still flying, the recovery point still exists, and the weather is bad enough to erode confidence without yet making the answer obvious. At that moment, is the autonomy trying to complete the mission, or is it trying to preserve a future in which the aircraft, the data and the organisation can still tell the truth about what happened?

Chapter 11

Ground Control Stations and Data Handling

What if the weakest part of an Arctic autonomous system is not the aircraft in the air but the small room, case, shelter or console from which people try to understand it? The question sounds unfair until a mission begins to depend on gloved hands, fogged displays, stiff controls, awkward posture, delayed data and a chain of digital records that must remain trustworthy long after the sortie is over. The aircraft may still be flying well. The control station may already be lying.

That possibility matters because ground control is often described too abstractly, as though it were simply the place where information arrives and commands depart. In severe cold, it becomes something harder and more consequential: the point where people, software, shelter, hardware, storage and procedure all have to agree quickly enough for the mission to remain legible. A cramped bench, a dim display, a touchscreen that assumes bare fingers, a cable arrangement that invites confusion, a removable drive whose provenance is uncertain, a data log that cannot later be reconciled with the aircraft's timeline - these are not side issues. They are the operating texture of reliability. The desk matters.

This is why ground control stations should be understood less as accessories to autonomy than as its constitutional centre. They determine what the human can see, what the human can question, what the system can preserve, and what the organisation can later prove about what happened. A machine in Arctic conditions is never only an aircraft or vehicle. It is also a chain of interfaces, files, version states, storage practices and agreements about who may plug in, update, carry, tag or interpret what. When that chain is strong, autonomy looks disciplined. When it is weak, the programme begins mistaking convenience for control and visibility for understanding.

The central tension lies there. Operators want fast access, responsive controls, flexible data movement and systems that cooperate easily across teams and partners. Harsh conditions demand something sterner:

interfaces that still work through cold gear and fatigue, cyber practices that protect reliability rather than merely decorate policy, records that remain auditable under weak bandwidth, and interoperability choices that admit the real cost of making different systems speak to one another. Ease and integrity are not always enemies, but they are not natural allies either. Every shortcut taken at the console eventually turns up somewhere else as delay, doubt, mistrust or preventable friction.

That tension matters because the ground station is where physical strain and informational strain become one problem. A pilot or operator in bulky cold-weather clothing cannot use controls, screens and menus as though seated in a dry laboratory. A mission lead making time-critical decisions from a fogging display or a cluttered shelter is not simply less comfortable. That person is operating with a different cognitive budget. Meanwhile the data systems behind the screen may already be carrying their own burdens: inconsistent tagging, uncertain version control, stale patches, removable media habits that weaken trust, and storage practices that make later review more difficult than the mission plan ever admitted. Human performance and data credibility therefore rise or fall together. A programme that separates them conceptually is already teaching itself the wrong lesson.

The stakes extend beyond individual missions. Ground control and data handling are where small compromises become institutional habits. A patch postponed for convenience can become a quiet source of configuration drift. A poorly governed transfer routine can weaken confidence in what data belongs to which sortie. A display layout that seemed tolerable in warm testing can become a safety issue once gloves, condensation and confined shelter space enter the picture. A coalition partner's format or frequency assumption can create a training and logistics burden that only appears after deployment, when the cost of incompatibility is no longer theoretical. These are durable problems because they do not merely produce isolated faults. They shape what the organisation comes to regard as normal.

That is also why credibility sits at the centre of this chapter. Autonomy programmes survive not only on performance claims but on whether their records can be trusted, their operators can function under actual conditions, and their systems can interoperate without constant informal repair by tired humans. A console is not credible because it looks modern. A data pipeline is not credible because it stores a lot. An interoperability promise is not credible because two vendors say their products are open. Credibility appears when the cold, the shelter, the gloves, the removable drive, the patch history and the mixed fleet all stop being background details and still fail to break the chain of understanding.

The path ahead follows that chain from the human hand on the interface to the institutional consequences of shared systems. It begins where cold weather makes simple control tasks slower, clumsier and more consequential than designers often admit. It then turns towards cyber hygiene, not as a separate technical priesthood but as part of everyday operational trust, before moving into the custody of data itself, where storage, tagging and retrieval begin deciding what can later be learned or defended. From there the argument widens into coalition and interoperability, where the promise of plug and play meets the colder realities of formats, frequencies, training and logistics.

A useful way to hold these questions together is to think of a filing cabinet in a storm. The drawers still open. The labels still exist. Papers can still be moved from hand to hand. But if the room is damp, the gloves are thick, the light is poor and the folders were filed carelessly to begin with, the cabinet stops being a tool of order and becomes another source of delay and error. Ground control stations work in much the same way. They do not merely display the mission. They organise the conditions under which the mission can still be understood, controlled and reconstructed.

The harder question, then, is not whether the aircraft can keep flying while the Arctic imposes its usual penalties. The sharper question is whether the people and systems on the ground can still see clearly, decide cleanly, preserve trustworthy records and cooperate honestly once cold gear, drifted configurations, intermittent bandwidth and mixed standards begin pressing on the console together - or whether the programme has built itself an elegant aircraft atop a desk that cannot carry the weight of its own claims.

Ergonomics, cold gear, and operator performance

On 28 November 1979, Air New Zealand Flight 901 flew into the slopes of Mount Erebus on Ross Island, Antarctica. The disaster belonged to crewed aviation, not to unmanned Arctic systems, and its causes were multiple. Yet one lesson from it remains brutally relevant to ground control design: severe environments do not simply challenge aircraft. They degrade the human act of seeing, interpreting and deciding. When the world outside becomes pale, ambiguous or deceptive, small flaws in the way information reaches a human operator stop being inconveniences and become part of the accident chain. The screen becomes terrain.

That is the right starting point for ergonomics in cold-weather control. People often hear the word ergonomics and imagine comfort, office furniture, preference or personal style. In Arctic operations, ergonomics is much less forgiving. Gloves thicken the hand and dull the fingertip. Fogged displays blur the very edge of a cue. Cramped shelters force awkward posture, poor sightlines and improvised tool placement. Cold reduces dexterity, patience and the speed with which a person can move from noticing a problem to acting on it. The result is not merely that the operator feels less comfortable. The result is that control accuracy and decision speed begin to change in ways the system may not have admitted in testing.

This is why human-machine interfaces in severe cold must be judged by their performance under actual constraints rather than by their apparent elegance in clean, warm demonstrations. Show the ordinary scene and the issue becomes painfully clear. A remote pilot or mission operator is standing or sitting in bulky winter gear, perhaps shoulder to shoulder with others, perhaps half-turned because shelter space is poor, trying to read a display that has begun to haze slightly from temperature transition. A menu requires fine selection. A map layer obscures a warning. A touchscreen assumes bare fingers or light gloves. A small acknowledgement button sits too close to a different command. Name the real problem and the engineering question sharpens: interface design under bodily impairment. Explain it and the implication follows. What looked like a tiny design choice in development now changes the chance of safe intervention in the field.

Gloves are a good example because they seem so mundane. Everyone knows people wear gloves in the cold. That superficial familiarity encourages bad design. A system is tested by someone slipping on a pair of light gloves for a few minutes, tapping through menus, nodding at the apparent usability, then returning to the old assumption that the interface is good enough. Real Arctic glove use is harsher. Thick gloves widen the finger, soften tactile discrimination, reduce pressure control and make repeated precise inputs slower and more fatiguing. If the operator removes the gloves to recover dexterity, then the interface has already forced a dangerous trade between control fidelity and bodily function. A control station that only works elegantly with warm bare hands is not ergonomically imperfect. It is conceptually false.

Fogged displays create a different but related falsehood. They do not necessarily render the screen unreadable. That is what makes them dangerous. A screen that has gone entirely opaque announces its failure.

A screen with softened contrast, edge haze, patchy condensation or glare from uneven shelter light merely taxes interpretation. Small icons become uncertain. Trend information takes longer to parse. Relative movement or subtle attitude cues become harder to trust. The operator squints, leans closer, shifts posture, wipes the display, rechecks a second pane, and all the while the mental cost of maintaining the same situational picture rises. That cost matters because severe operations rarely fail only through one large mistake. They fail through the quiet accumulation of time lost, confidence thinned and attention diverted.

A useful historical example lies in the 2009 loss of Air France Flight 447 over the Atlantic. The environment was different, and no one should force a simplistic analogy. Yet the structural lesson is valuable: once reliable cross-checks weaken and information must be interpreted under stress through instruments and system behaviour alone, the design of those cues and the operator's ability to read them become central to survivability. Arctic remote operations push people into a related condition, though from the ground rather than from the cockpit. When visibility is poor and the aircraft is distant, the interface becomes the operator's weather window, instrument panel and decision theatre all at once. If that interface has not been proven under cold-induced bodily limitations, the organisation is gambling on a version of the human user who will not actually be present.

A second example can be taken from publicly discussed polar and remote field operations, including scientific and survey work, where teams repeatedly discover that shelter layout and operator posture alter performance as much as software quality does. A cramped enclosure changes where people place elbows, how they hand over controls, whether one display blocks another, whether condensation accumulates differently across screens, and whether one operator's body obstructs another's line of sight during a critical phase. These details sound pedestrian. They are not. A poor shelter can turn a well-designed interface into a clumsy one by forcing the body into postures for which the interface was never really meant. The error is then blamed on field conditions, as though field conditions were not supposed to be the conditions of use.

This is why ergonomics must be treated as part of reliability, not as a user-experience afterthought. Reliability in harsh operations depends not only on whether the aircraft receives the correct command, but on whether the operator can issue that command in time, under the actual physical penalties of the environment, without accumulating hidden hesitation or ambiguity. A control that takes one extra second to find, confirm or execute may be trivial in a lab. It is not trivial during a gusty recovery or a degraded-link abort. A cluttered display that forces an extra cognitive step may be acceptable in briefing. It is not acceptable when the operator is already balancing a payload feed, a route deviation, a battery margin and a rising sense that the landing point no longer looks honest. Human performance is a systems variable.

Still, a fair limitation belongs here. Not every cold-weather interface challenge can be designed away entirely. Some missions will always impose awkward posture, imperfect shelter conditions or compromised dexterity. Portable systems cannot all become spacious, heated command rooms. Some operators also adapt impressively, developing rhythms and workarounds that preserve effectiveness under difficult circumstances. It would be careless to imply that every cramped shelter or fogged screen immediately disqualifies the operation. Humans are resourceful, and good teams often keep bad designs alive longer than they deserve.

But that limitation only sharpens the main point. Because adaptation and workarounds exist, organisations become tempted to mistake them for design adequacy. A team that repeatedly succeeds despite awkward inputs, poor display readability or cramped layout is often masking an ergonomics debt the system still owes. That debt will eventually come due when a less experienced operator takes the watch, when weather worsens, when a second anomaly arrives at the same time, or when the same awkward shelter must support a longer and more cognitively expensive sequence than usual. The reconciliation is therefore stern. Respect operator ingenuity, but do not let it flatter weak interface design.

This has consequences for testing culture. An interface proven by a technician in shirtsleeves on a bench tells the organisation very little about how it will behave in real Arctic service. Proper testing must include gloves that represent actual use, condensation and glare conditions that resemble actual shelter transitions, seating or standing arrangements that match field layout, and task sequences long enough to show how accuracy and speed degrade when cold discomfort, fatigue and attention pressure accumulate together. Anything less is not really interface testing. It is demonstration. A system that has never been trialled under the bodily facts of cold weather does not yet know what sort of human it depends on.

The quiet motif here is the windowpane. A clear window seems neutral until it fogs, frosts or reflects, and then suddenly one notices how many judgements had been passing through it without effort. Ground control interfaces work in much the same way. When they are clean, legible and physically usable, operators treat them as transparent. In the Arctic they often stop being transparent. Every layer of bodily interference - glove, haze, cramped angle, awkward reach, cold stiffness - makes the pane thicker. The operator still sees through it, but not for free.

This is why simple interface choices can become safety-critical. Font size, button spacing, confirmation logic, control travel, display brightness, physical separation between dissimilar commands, the ability to acknowledge a warning without obscuring the underlying flight state, the location of critical information relative to the operator's natural line of sight - none of these is glamorous. Yet each determines whether the system can still be run truthfully when the environment begins reducing the operator to something slower, stiffer and more cognitively taxed than the design team may have imagined. Cold-weather reliability begins to fail at the point where the interface demands a body and a mind that the shelter cannot actually provide.

The unresolved question, then, is not whether an operator can eventually make the right decision from a cold, cramped control station. The harder question is whether the interface still permits accurate, timely and confident action before the penalties of gloves, fogged screens and bad posture quietly consume the margin the mission was counting on all along.

Cyber hygiene as a reliability issue

In June 2010, the discovery of Stuxnet forced a wider public to confront something engineers and operators had long known in quieter ways: digital systems do not fail only through dramatic sabotage or spectacular intrusion. They also fail through the patient corruption of trust. A controller still runs, a display still updates, a file still opens, a device still appears to belong to the right network, and yet the system has already become less believable than it looks. The software still answers.

That lesson matters sharply for Arctic autonomous operations because cyber hygiene is too often discussed as a specialised security concern hovering somewhere above the "real" business of flying, recovering and maintaining vehicles. In practice it belongs much closer to the workbench and the launch line. A ground station with uncertain patch status, a support laptop drifting from the approved configuration, a removable drive of unclear provenance, or a mismatched software image loaded under schedule pressure can all degrade reliability long before anyone uses the word cyber. The aircraft may still launch. The question is whether the organisation still knows what system it has actually launched.

This is why cyber hygiene should be treated as an issue of operational robustness before it is treated as a matter of defensive prestige. Show the ordinary field rhythm and the point becomes practical at once. A team updates one control station and not another because the weather window is narrow. A sensor workstation retains an old dependency because a new one has not yet been verified against mission software. An operator moves data on removable media because bandwidth is intermittent and time is short. A support machine accumulates small local workarounds through the season until it is no longer quite the machine the official configuration claims it is. Name the problem and the argument sharpens immediately: digital drift under operational pressure. Explain it and the implication follows. Reliability begins to fray not when the system stops functioning, but when nobody can say with full confidence which version of the system is presently in charge.

Patching sits at the centre of this tension because it appears so straightforward from a distance and becomes so morally awkward in live operations. Everyone understands the principle that known defects, vulnerabilities and software weaknesses should not be left to age untreated. Yet patching also changes systems, and changing systems in harsh operations can feel like introducing fresh uncertainty into a machine that must already cope with cold, poor visibility, intermittent communications and strained maintenance cycles. This is why organisations slip into bad habits at both extremes. Some defer patches until the field season is over, allowing ageing software to become part of the environment. Others patch opportunistically, accepting uncoordinated variation across consoles, laptops, payload tools and storage systems. Both habits degrade trust. One preserves known weakness. The other undermines configuration coherence.

That coherence matters because availability depends on repeatability. A ground control station is not reliable merely because it powers on and connects. It is reliable because the team knows how it will behave relative to the aircraft, the payload, the data pipeline and the other machines around it. Configuration drift attacks that knowledge quietly. One workstation displays a field differently. Another handles storage timestamps differently. A third no longer synchronises or validates in the same way after an untracked software change. None of these differences need look like obvious security incidents to become operationally significant. In the Arctic, where environmental ambiguity already taxes attention and time, hidden digital inconsistency is especially corrosive. It turns troubleshooting into guesswork and post-mission analysis into negotiation.

A useful historical example lies in aviation maintenance and software control more broadly. Modern aircraft systems are managed under strict configuration disciplines not because engineers are pedantic by nature, but because once systems become software-mediated, small undocumented differences can produce large interpretive confusion. The relevance to unmanned Arctic systems is direct. A control station, mission laptop, payload processor and logging tool do not need to be maliciously altered to become unsafe partners. They need only diverge just enough that the operator can no longer rely on common behaviour across the fleet. High-level cyber hygiene is therefore less about drama than about preserving sameness where sameness is the basis of trust.

A second example comes from remote scientific and expeditionary operations, where teams often rely on removable media, local copies and seasonal workarounds because bandwidth is scarce and infrastructure is thin. Publicly discussed field programmes repeatedly reveal the same structural risk: the very tools that preserve mission continuity can also weaken confidence in data provenance and system state if they are not governed with unusual care. The lesson worth borrowing is not one of fear. It is one of respect for how easily "temporary" digital habits become permanent operational assumptions. In a harsh theatre, the improvised file transfer or unofficial laptop image is not just a convenience. It becomes part of the system the mission is trusting.

This is why removable media policy belongs inside reliability culture rather than off to one side as a specialised security ritual. The issue is not only whether a storage device carries obvious malicious code. The issue is whether a programme can still trust the lineage of what was moved, when it was moved, on which machine it was opened, and how it may have altered the state of the systems it touched. In bandwidth-poor conditions, removable media often feels unavoidable. That does not make it trivial. It means the organisation must treat such media as operationally consequential objects, because they affect not just security posture but auditability, software consistency and the credibility of mission records. A stray drive can compromise trust more quietly than a broken sensor.

Still, a fair limitation must be admitted. Not every instance of imperfect cyber hygiene leads to failure, and not every field workaround is a sign of institutional decay. Harsh operations sometimes require pragmatic decisions to keep missions moving, especially when infrastructure is sparse and the costs of delay are real. Some systems can tolerate limited asynchrony or deferred patching better than others. It would be foolish to imply that any deviation from digital ideality instantly disqualifies a programme from serious work.

But that limitation only clarifies the standard. The question is not whether expedient choices will ever be made. The question is whether the organisation knows when such choices have crossed from bounded pragmatism into hidden unreliability. A mature programme records the deviation, contains its scope, understands its operational meaning and restores coherence deliberately. An immature one lets small exceptions accumulate until the fleet is running on a folklore of local fixes and nobody can distinguish ordinary behaviour from digital accident. The reconciliation is stern but useful: cyber hygiene should support continuity without turning continuity into an excuse for uncertainty about the system's own identity.

This becomes especially important when incidents occur. A platform behaves oddly on approach. A payload record does not match the expected timeline. A mission log appears incomplete or inconsistent across storage locations. A control station refuses a command sequence another station handled normally the day before. At such moments, the organisation needs to know whether it is confronting weather, operator strain, hardware degradation or digital incoherence. Poor cyber hygiene makes that distinction much harder. It does not merely create risk in advance. It weakens diagnosis afterwards. The cost is paid in slower learning, wider blame and thinner trust.

The quiet motif here is the keyring. A keyring looks harmless enough: small, useful, familiar, easily passed from hand to hand. Over time it accumulates duplicates, unknown keys, worn tags and keys that once opened something important but no one now quite remembers. Digital access, storage devices, local images and seasonal workarounds often behave in the same way. They remain manageable until their very familiarity hides the fact that the organisation is carrying more uncertainty than it knows. Cyber hygiene is the discipline of refusing that drift before a routine object becomes a reliability hazard.

This is why safety culture and digital discipline belong to the same conversation. A programme that cannot keep its software baselines, data movement and removable media habits legible will eventually struggle to keep its maintenance history, incident review and operational confidence legible as well. The failure may first appear as a delayed patch, a mismatched file, a stubborn workstation or a questionable log. Later it becomes something larger: a fleet whose behaviour can no longer be separated cleanly into environmental difficulty, human error, hardware weakness and digital inconsistency. At that point, reliability has already become an argument rather than a property.

The unresolved question, then, is not whether cyber practice matters in an abstract strategic sense. The harder question is whether the programme's daily digital habits still allow operators and maintainers to trust what kind of system they are actually using when the weather is poor, the bandwidth thin and the mission under pressure - or whether the aircraft's most serious cold-weather weakness has quietly migrated into the laptops, drives and version states on the ground.

Data pipelines, storage, and auditability

On 1 June 2009, Air France Flight 447 disappeared over the Atlantic between Rio de Janeiro and Paris. For a long time, the world knew far less about what had happened than it wanted to know. The aircraft was gone, the event had ended, and yet the decisive struggle moved elsewhere: into records, timestamps, recovered devices, reconstruction and the credibility of data that could still speak after the flight could not. The lesson is plain. A system does not become intelligible merely because it once produced data.

That lesson reaches directly into Arctic autonomy. People often imagine sensor data as a kind of natural harvest: the aircraft flies, the payload sees, the logs accumulate, and the organisation later analyses what was gathered. Severe conditions make that picture dangerously naive. In the High North, bandwidth is often intermittent, retrieval delayed, storage improvised, and the difference between what was sensed, what was transmitted, what was tagged correctly, and what can later be trusted may grow wider with every weak link in the chain. The record is not the mission's shadow. It is another operational product, and it must be built with the same seriousness as the aircraft's route or recovery plan.

This is why data pipelines deserve to be understood as part of the vehicle's control architecture rather than as a quiet administrative service running in the background. Show the ordinary field problem and the stakes become immediate. A platform collects imagery, thermal scenes, state estimates, health data, timing signals, operator inputs and mode changes while crossing a region where communications are unstable and storage must absorb what the link cannot carry in real time. Later, someone asks a question that matters: what did the aircraft know, what did the operator see, what happened first, and which machine state deserves belief? Name the issue properly and the engineering problem sharpens: auditability under degraded transfer. Explain it and the implication becomes severe. If the data pipeline has not preserved provenance, timing and context as carefully as it preserved raw content, then the programme may possess terabytes of material and still lack trustworthy knowledge.

That is why storage is not merely a capacity question. A system can store a great deal and still store it badly. Data that arrives without clear tagging, synchronisation, environmental context or configuration lineage becomes hard to compare, hard to retrieve and, in the most consequential moments, hard to defend. The image may still exist. The question is whether anyone can later say with confidence which sensor produced it, under what system state, with what timing relative to aircraft manoeuvre, operator command, navigation confidence or communications degradation. In severe operations, this is not an archival nicety. It is the basis on which incident review, mission assessment and model improvement either become serious or collapse into impression.

A useful historical example lies in aviation accident investigation more broadly, where the value of flight recorders has never rested on the sheer volume of captured information. Their importance lies in disciplined timing, preserved sequence and the ability to reconstruct a coherent chain from cockpit inputs to aircraft behaviour. The relevance to autonomous systems is exact. A drone, remote vessel or robotic platform should not merely collect sensor products and health logs; it should preserve them in a relationship strict enough that later analysis can answer not only what happened, but what the system believed was happening as events unfolded. The distinction matters because many operational failures are not simple hardware breakages. They are failures of interpretation, timing and confidence. Weak records do not merely obscure blame. They obscure causality.

A second example comes from planetary robotics and remote science operations, where delayed transmission and limited bandwidth have long forced organisations to decide what is sent first, what is buffered, what is compressed, and what metadata must never be stripped away if the mission is to remain scientifically and operationally honest. The important lesson is not one of exotic distance. It is one of hierarchy. Not all data is equal in the moment, and not all data is equally valuable later. A mature programme knows the difference between live information needed for immediate authority, records needed for incident reconstruction, and higher-volume products that can tolerate delay so long as their lineage remains intact. The Arctic punishes programmes that do not make these distinctions early, because it gives them many opportunities to lose context while still believing they have preserved substance.

This is also why tagging becomes more consequential than many engineers or operators first expect. Tagging sounds bureaucratic until one imagines trying to reconstruct a difficult recovery in gusting wind, a navigation anomaly in poor visibility, or a disagreement between two data products after a mission conducted with intermittent links. A file called by its time alone may be insufficient. A payload clip detached from aircraft mode, sensor state or location confidence may be less useful than a smaller, better-described record. Tags are not labels pasted onto an otherwise complete truth. They are the structure that lets later investigators, maintainers and analysts recover the relationship among event, machine and environment. Without them, retrieval becomes archaeology.

Still, a fair limitation must be admitted. No fielded programme can preserve everything in perfect fidelity, with perfect synchronisation, under all operating conditions. Bandwidth will fail. Storage will be finite. Compression and prioritisation will remain unavoidable. Some records will be noisier, thinner or later than anyone would prefer. It would be foolish to pretend otherwise. The question is not whether the data pipeline can be made immaculate. The question is whether it remains honest about what may be trusted, what has been degraded, what can be correlated, and where uncertainty has entered the record itself. A system that confesses those boundaries is far more useful than one that produces an ocean of files whose credibility is taken for granted and therefore never properly tested.

That honesty matters because autonomy programmes increasingly depend on their records not only to explain past sorties but to shape future behaviour. Model tuning, perception refinement, maintenance planning, threshold adjustment and doctrine revision all lean on stored data. If that corpus is inconsistent, poorly tagged or weakly auditable, the programme begins teaching itself from contaminated memory. It may reinforce the wrong pattern, miss the real precursor to failure, or mistake a storage artefact for an operational truth. The danger is not merely that one incident will be harder to explain. The deeper danger is that an entire organisation begins improving against an unstable picture of its own experience. A bad record does not simply fail history. It poisons adaptation.

This is why auditability is such a stern and useful word. It asks not whether the programme has data, but whether the data can sustain scrutiny. Can a difficult event be reconstructed across aircraft logs, payload products, operator inputs and communications states without resorting to intuition or wishful stitching? Can a maintenance anomaly be traced back to the right environmental exposure and system configuration? Can two seasons of operations be compared without discovering too late that naming, timing or metadata conventions changed so much that the apparent trend means little? Auditability is the point where storage discipline becomes institutional maturity. A programme that cannot audit itself cannot really learn itself.

The quiet motif here is the thread through beads. A handful of beads in a box still proves that beads exist. It does not make a necklace. Data files scattered across machines, drives, temporary folders and partial links behave in much the same way. The mission may have produced many valuable fragments, yet without the thread of timing, tagging and configuration coherence they cannot carry the weight later placed upon them. The thread is what makes retrieval more than rummaging.

That is why autonomy programmes rise or fall on the credibility of their records. The aircraft may perform impressively, the operators may act with discipline, the payload may gather valuable scenes, and the sustainment team may preserve the machine through punishing conditions. But if the record cannot later say clearly what happened, what was known, what was changed, and what deserves trust, then the programme's apparent sophistication will eventually hit a hard limit. It will struggle to explain incidents, to defend decisions, to improve models, and to distinguish genuine learning from institutional folklore.

The unresolved test comes after the sortie, when the aircraft is back or perhaps not, the weather has moved on, and the organisation must decide what the mission really taught. At that moment, are the records strong enough to let the programme confront the truth of its own system, or has the pipeline quietly transformed hard-won experience into a pile of files that look rich until someone asks the first serious question?

Interoperability and coalition friction

In 1941, when Allied forces began operating under the Arctic convoy system to Murmansk and Archangel, they discovered a truth that has outlived the ships, radios and staff work of that era: cooperation is never created by shared intent alone. It depends on whether different organisations can actually exchange signals, interpret instructions, trust records and sustain one another without translating every move into a local dialect first. The ice did not care whose flag a vessel flew. The friction remained human and technical.

That truth returns, sharpened, in modern autonomous operations. People often speak of interoperability as though it were a natural dividend of contemporary technology. If systems are digital, networked and nominally standards-based, then surely they should cooperate more easily than earlier generations of equipment. In the Arctic, that hope meets cold reality very quickly. Different standards, frequencies, encryption arrangements, file structures, metadata conventions, maintenance assumptions and operator interfaces do not merely create annoyance. They shape what can be launched, what can be shared, what can be trusted and how much human labour must be spent turning one partner's "common format" into something another partner can actually use. Plug and play is a phrase. Winter is an audit.

This is why interoperability should be treated not as a technical feature but as an operational bargain. Show the practical scene and the problem becomes unmistakable. One team arrives with aircraft that speak fluently to its own control station, log data in its own structure, use its own support devices and assume its own spectrum arrangements. Another team brings a different set of systems, equally coherent inside their home architecture. On a planning slide, the combined force looks richer, more resilient and more flexible than either element alone. On the ground, someone must now reconcile interfaces, connector standards, file naming, training assumptions, data transfer habits, mission planning tools, environmental settings and the plain fact that an operator already stressed by cold and wind cannot also become a full-time translator between ecosystems. Name the issue and the argument sharpens at once: coalition friction is logistics in disguise. Explain it, and the implication follows. A system may be individually sophisticated and still operationally clumsy in company.

Frequencies make this plain because they remind everyone that cooperation begins in physics before it reaches policy. If systems rely on different bands, antenna arrangements or communications assumptions, then shared operation is never simply a matter of goodwill. It becomes a question of deconfliction, coverage, equipment carriage and who must adapt to whom when the weather is already degrading signal quality. A platform that is perfectly credible within its own communications architecture may become burdensome in coalition use because the local support burden rises: more hardware to carry, more spectrum management to perform, more chances for misconfiguration, more points at which a weak or delayed link is blamed on "the environment" when the deeper issue is architectural mismatch. The Arctic does not forgive this easily because its communications margins are already thin.

Data formats create a different, subtler strain. Shared operations often assume that because files can be exported, they can be integrated. That is rarely the whole truth. A sensor product moved from one system to another may lose metadata, timing coherence, location confidence or the contextual relationships that made it operationally meaningful in the first place. A receiving team may technically ingest the file and still not be able to use it without manual cleaning, reinterpretation or delayed verification. That labour matters because it usually falls on exactly the people already carrying weather, maintenance and mission pressure. Interoperability, in other words, is not proven when data can be handed over. It is proven when the receiving organisation can act on it without rebuilding the meaning by hand.

A useful historical example lies in NATO operations more broadly, where coalition work has long depended on a mixture of formal standardisation and relentless practical adaptation. The lesson worth keeping is not triumphal. It is cautionary. Even among close partners with long habits of cooperation, shared operation still demands continual work on doctrine, interfaces, language and procedure. The Arctic magnifies that work because the theatre adds distance, sparse infrastructure, cold-weather logistics and harsh communications conditions to the usual burdens of multinational coordination. Any programme that imagines interoperability as a procurement checkbox rather than a continuing sustainment problem is already behind the environment.

A second example can be drawn from polar science and remote civil operations, where organisations frequently discover that "common data" and "shared systems" conceal large differences in how observations are tagged, moved, archived and trusted. Publicly described collaborative campaigns often succeed not because everyone used identical tools, but because the participants invested serious labour in agreeing what would count as the same measurement, the same timestamp, the same place, the same version, the same condition of validity. That lesson translates directly to autonomous systems. Common purpose does not remove the need for common definitions. If anything, it increases it, because the number of quiet mismatches rises with every additional partner and platform.

This is why training burden should be counted as part of interoperability rather than as a downstream inconvenience. A system that can be used safely only by its home team is not truly easy to integrate, however open its interfaces may appear. If coalition partners must learn a new visual language, a new menu logic, a new maintenance rhythm, a new storage convention or a new set of exception-handling rituals simply to cooperate around one platform, then that platform is importing cost into every mission. Those costs are often hidden because they appear as familiarisation, adaptation or minor local workaround. In harsh environments, such burdens become more serious. Cold strips away spare attention. Cramped shelters reduce the room for ad hoc teaching. Fatigue makes translation slower and mistakes harder to catch. A force that looks interoperable in briefing can become educationally overloaded in the field.

Still, a fair limitation should be admitted. Full uniformity is rarely possible and not always desirable. Different partners may have legitimate reasons for keeping distinct systems, standards or security practices. Diversity can even provide resilience if one architecture fails where another survives. It would be simplistic to suggest that coalition operations require everyone to converge on one platform, one format or one frequency plan before meaningful cooperation can occur. Real operations often succeed through partial compatibility, disciplined liaison and a mixture of standardisation and intelligent tolerance.

But that limitation only clarifies the real standard. The question is not whether systems are identical. It is whether the remaining differences are understood, bounded and supportable under actual Arctic conditions. A mature coalition accepts that some translation and adaptation will remain necessary, then asks who performs it, with what tools, under what weather and shelter constraints, and with what consequences for speed, trust and accountability. An immature one keeps repeating "plug and play" until the first difficult sortie reveals that the plug fits only after three adapters, two local scripts, one extra laptop and a patient operator who already had too much to do.

This is why scepticism about interoperability claims is not cynicism. It is operational hygiene. The serious reader should ask awkward questions. Does shared operation survive intermittent bandwidth, or only ideal connectivity? Do records remain auditable after passing between systems, or merely visible? Can one partner maintain another's equipment at the level actually required by the mission, or only at the level required by a demonstration? Does training produce fluency, or just a veneer of recognition? Does a common picture still mean the same thing once metadata, confidence and timing have travelled through several architectures? These questions reveal whether the coalition is building a real combined capability or merely arranging adjacent islands of competence.

The quiet motif here is the adaptor plug. On a desk, it feels trivial, even reassuring: a clever little object that lets one device speak to another. In the field, a chain of such adaptors becomes a warning sign. Each extra conversion step may work, but each also imports fragility, delay, failure points and new opportunities for misunderstanding. Coalition interoperability often behaves in just this way. One adaptor is practicality. Too many indicate that the system was never as naturally compatible as the briefing implied.

This matters because Arctic logistics are already burdened enough without importing avoidable translation work into every mission. Extra cables, extra support devices, extra conversion routines, extra training layers, extra storage handling, extra uncertainty about what exactly was shared and in what condition - all of these become forms of friction that the environment then magnifies. Cold makes equipment less forgiving. Distance makes resupply slower. High winds and poor visibility make time more valuable. Under those conditions, interoperability choices shape not only cooperation but endurance. A coalition can be well intentioned and still mechanically exhausted by its own interfaces.

The unresolved question, then, is not whether different systems can be made to exchange something in principle. The sharper question is whether they can cooperate under Arctic pressure without turning every shared mission into a hidden exercise in translation, extra carriage and improvised trust - and whether the promise of plug and play still survives once the gloves are on, the shelter is cramped and the weather has stopped tolerating polite technological fiction.

Chapter 12

Counter-Drone Threats and Electronic Attack

What happens when a system already weakened by cold, distance and poor reference enters a contest against someone who understands those weaknesses and chooses to push on exactly the same seams? That is the Arctic problem in its hardest form. Electronic attack sounds, at first, like a separate layer of danger added on top of the weather. In reality the distinction is often false. The environment has already been denying, distorting and thinning the margins on which the machine depends. An adversary merely learns to speak the same language more deliberately.

That is why the Arctic should not be imagined as a neutral stage onto which hostile action later arrives. It is already a denial system of remarkable discipline. Signals fade oddly. Surfaces confuse perception. Low contrast weakens confidence. Recovery windows narrow. Small navigation errors become larger moral problems because alternatives are sparse and timing is expensive. A drone that appears competent in easier theatres may therefore enter the High North already carrying hidden dependency: on clean sensing, on trustworthy position, on uninterrupted control logic, on the operator's ability to distinguish ordinary degradation from something more purposeful. The battery weakens. The picture whitens. The machine begins to ask for faith.

That faith is exactly what a serious adversary will try to tax. Jamming, spoofing and deception matter not because they conjure entirely new physics, but because they exploit a machine and an organisation already living near the edge of certainty. A false cue is more powerful where the real cue was never generous to begin with. A delayed response is more dangerous where recovery space is already thin. A misleading signature matters more where snow, ice, sea clutter and shifting backgrounds have already made identification expensive. The contested Arctic is therefore not a place where natural degradation stops and hostile interference begins.

It is a place where the two can blur until failure itself becomes difficult to classify.

The central tension lies there. Organisations want unmanned systems because they promise persistence, reach and reduced exposure for human crews. Yet those same systems often rely on layers of sensing, navigation and communication that the Arctic degrades before any adversary acts. The more a programme depends on those layers, the more attractive it becomes to anyone able to disturb them further. This is not a technical paradox to be solved once and filed away. It is a recurring strategic bargain. Every gain in stand-off, automation or distributed presence must be weighed against the question of how the machine behaves when its confidence becomes contested and when the organisation around it can no longer pretend that uncertainty is accidental.

The stakes are therefore larger than the survival of a single airframe. When electronic attack and environmental denial combine, they challenge the whole credibility of the programme. Can the system still distinguish bad weather from bad intent? Can operators still interpret loss of confidence without rushing into comforting but false explanations? Can the organisation preserve enough discipline that a stressed machine does not turn one failed recovery, one confused signature or one ambiguous link event into a broader crisis of trust? A force that cannot answer those questions cleanly will begin to oscillate between two equally poor habits: reckless confidence when the machine should retreat, and paralysing suspicion when the machine still has useful margin. Both habits waste capability. Both teach the wrong lessons.

That is why counter-drone thought in the Arctic must remain conceptually stern and operationally modest. The subject invites theatrical language because it sits close to conflict, secrecy and technological bravado. But the most important truths here are often duller. Resilience depends on disciplined assumptions about failure. Signature management begins with understanding that every platform is visible in more than one way, and often not in the way its designers found most flattering. Escalation control is less about bold response than about deciding, in advance, what losses are acceptable, what behaviour the autonomy may adopt under pressure, and when recovery attempts stop being a mark of seriousness and start becoming a refusal to admit that the conditions have changed. The strongest systems are not the ones that promise invulnerability. They are the ones that stay governable when invulnerability proves imaginary.

The path ahead follows that governability through increasingly contested forms of dependence. It begins with the uncomfortable recognition that the environment and the adversary may be attacking the same weak points at once, then turns to the problem of detection and signature, where cold air, cluttered backgrounds and odd atmospheric effects complicate the business of knowing what is present and what it means. From there the argument moves into resilience, not as a shopping list of hardware virtues but as a discipline joining redundancy, diversity and trained modes of operation, before tightening around escalation and risk, where organisations must decide what they are truly willing to lose and what kinds of autonomous behaviour they will or will not authorise once pressure begins to mount.

A useful way to hold these questions together is to think of a cracked windscreen in freezing rain. The vehicle still moves. The road may still be there. But every new distortion joins one already in progress, and the danger lies not only in each crack by itself but in the way they multiply the driver's uncertainty about what is real, what is artefact and how long continued motion remains honest. Counter-drone pressure in the Arctic works in much the same way. Environmental distortion and hostile distortion need not be identical to become mutually useful.

The harder question, then, is not whether a drone can survive some amount of interference or endure some amount of weather. Many can. The sharper question is whether the organisation has drawn clear enough limits around confidence, behaviour and acceptable loss that, when cold air and hostile pressure begin bending the same signals, the machine still fails in a way that preserves truth rather than merely preserving motion for a little longer.

Environmental and adversarial denial combined

In October 1962, during the Cuban Missile Crisis, one of the least sentimental facts of strategy became impossible to ignore: systems under pressure do not fail only because an enemy attacks them. They fail because fear, uncertainty, degraded information and contested signals begin to work together faster than institutions can interpret them. The famous images are of missiles and ships. The deeper lesson is about ambiguity. A stressed system becomes easier to mislead.

That lesson belongs in the Arctic with unusual force because the environment is already doing part of the adversary's work before any hostile actor lifts a finger. Signals thin. Visual reference weakens. Low contrast obscures motion and intent. Ice, sea clutter and shifting light complicate what sensors believe they are seeing. Wind, cold and recovery geometry reduce the margin available for correction once something goes wrong. A drone operating in these conditions is not entering neutral air. It is entering a theatre that is already denying, distorting and delaying. The static is natural first.

That is why it helps to think of the Arctic as a pre-existing denial system. Show the practical case and the point sharpens at once. A platform relies on satellite navigation, radio links, optical cues and stable control assumptions to move safely and return credibly. The environment degrades each of those in its own way: intermittent visibility, uncertain horizons, reflective or texture-poor surfaces, awkward propagation, cold-driven power penalties, gusts near terrain or ships, sparse recovery options. Name the governing idea and the argument becomes clearer: environmental denial is not metaphor. It is the cumulative weakening of dependencies on which autonomy rests. Explain it and the implication follows. An adversary entering this picture does not need to create confusion from nothing. It needs only to amplify weakness already present.

This is where jamming, spoofing and deception become more dangerous than they would be in a more forgiving theatre. Not because the technologies are magically transformed by latitude, but because the target system is already living close to the edge of confidence. A false positional cue matters more when terrain is feature-poor and visual correction is scarce. A degraded link matters more when recovery sites are few and weather is already tightening the timetable. A deceptive signal or ambiguous contact matters more when the background is already cluttered with ice, shadow, glare or atmospheric oddity. In easier conditions, the machine and its operators might cross-check their way back towards certainty. In the Arctic, they often have less to cross-check with. The lie lands on tired ground.

This is why failure modes should be seen as contested terrain rather than as tidy, isolated malfunctions. A drone drifting off a route in clear temperate conditions may invite a straightforward diagnosis: navigation fault, operator error, unexpected wind, perhaps a software issue. In the High North, the same drift can sit at the intersection of several degrading truths at once. GNSS confidence narrows. Visual features repeat. Communications wobble. Low temperatures alter battery behaviour and thus the energy margin available for correction. Then an adversary adds noise, false cues or targeted interference. The resulting failure is not merely technical. It is a fight over interpretation. If the organisation treats these layers as separate in principle, it will understand the event too slowly in practice.

A useful historical example lies in the public understanding of satellite navigation denial more broadly. Even without going into operational details, modern users of navigation-dependent systems have repeatedly learned that disruption does not need to destroy every signal to become effective. It only needs to make trust more expensive than the mission can afford. That lesson is magnified in the Arctic because navigation was never cheap there to begin with. The environment had already raised the price. The adversary simply notices the inflated bill. A second example can be taken from broader electronic warfare history, where deceptive effects often succeed not by overwhelming every system at once but by exploiting moments when humans and machines are already struggling to tell nuisance from threat. In Arctic operations, those moments are plentiful because natural degradation creates a constant background of plausible excuses.

Still, a fair limitation must be admitted. Not every anomaly in the Arctic is adversarial, and not every degraded signal or odd recovery is evidence of hostile action. That is precisely what makes the theatre so intellectually demanding. Organisations can drift into paranoia as easily as they can drift into complacency. A system that attributes too much to the enemy will misread natural failure, squander confidence and perhaps invite needlessly escalatory responses in doctrine or planning. A system that attributes too little will normalise hostile pressure as "just the weather" until the pattern is much harder to reverse. The problem is not merely technical discrimination. It is organisational judgement under persistent ambiguity.

That is why the deeper issue is dependence, not attack in the abstract. A strong programme does not ask only whether its platforms can survive direct interference. It asks which assumptions about navigation, timing, sensing, bandwidth, recovery and operator interpretation become dangerously expensive when the environment and the adversary begin pressing on the same weak points. Some dependencies are obvious and therefore easier to protect. Others are more cultural. Crews become accustomed to marginal links. Operators learn to accept vague imagery because the weather has often made it vague before. Maintainers begin treating intermittent anomalies as part of polar life. These habits are understandable. They also create a soft landing zone for hostile amplification. The adversary's best tool may sometimes be the organisation's own prior acclimatisation to ambiguity.

This is where resilience begins to look less like hardening everything and more like teaching the system to stay honest when certainty is being stripped away. Honesty, here, means recognising that a contested Arctic environment can make machine confidence appear steadier than it deserves to be. It means refusing to let a platform continue under the fiction that environmental degradation and hostile degradation can always be cleanly distinguished in real time. It means accepting that some forms of failure cannot be neatly labelled at the moment they occur, and that doctrine must therefore narrow authority early rather than waiting for courtroom proof. This is not timidity. It is a serious response to a theatre in which the environment and the enemy can collaborate without ever formally meeting.

The quiet motif here is the cracked window. One crack may still leave enough view to continue. A second, from a different angle, changes the whole act of looking. Soon the eye is working harder than the mind admits, and every decision is being made through a surface whose damage is cumulative rather than singular. Arctic denial works in much the same way. So does adversarial exploitation. The machine may still move, the feed may still arrive, the control link may still respond, yet the conditions of trust have already been fractured in more than one place at once.

This is why contested autonomy in the High North cannot be reduced to a contest between a clever attacker and a well-engineered drone. It is a contest over the credibility of the entire chain: sensors, operators, links, navigation, recovery logic, maintenance history and institutional habits of explanation. The environment is already eroding that chain. Hostile action merely chooses where to press harder. An organisation that sees these as separate worlds will continue speaking the language of accident when it should be speaking the language of compounded pressure. An organisation that collapses everything into enemy action will lose its grip on the ordinary, costly physics of Arctic operations. The mature position sits uneasily between those errors and stays there.

The hardest question, then, is not whether the next failure will be natural or man-made in some clean categorical sense. The sharper question is whether the system can still recognise when its dependencies have become so contested - by weather, weak reference, noisy signals, deceptive cues or some blend of all four - that continued confidence has stopped being evidence-based and started becoming a form of denial in its own right.

Detection, tracking, and signatures in cold air

In 1935, when Robert Watson-Watt demonstrated early practical radar work in Britain, the important lesson was not merely that objects in the sky could be detected at distance. It was that visibility had ceased to belong to one sense alone. The eye could fail, the weather could interfere, the horizon could deceive, and yet another domain might still reveal what the first had missed. That shift matters in the Arctic because the environment is unusually good at punishing any observer who believes one kind of signature is the whole truth. The sky is never just visual.

That is the right place to begin thinking about detection and tracking in cold air. People often talk about a platform as though it were quiet or visible, hidden or conspicuous, easy to find or hard to find. In the Arctic those opposites rarely stay tidy for long. A drone may be acoustically modest over open water and visually obvious against flat snow. It may be hard to spot by eye in haze and far easier to notice in thermal contrast at the wrong moment. It may emit little heat at one phase of flight and declare itself electromagnetically through the system needed to control or recover it. Detection is therefore not a single contest between object and observer. It is an argument among domains, backgrounds and changing conditions.

This matters because cold air does not simply make the Arctic a cleaner stage on which signatures stand out neatly. Show the practical case and the problem becomes immediately more difficult. A small aircraft crosses over sea ice, low cloud and broken glare. The eye sees little except tonal sameness. Acoustic detection may be altered by wind and surface effects. Infrared contrast may sharpen or soften depending on what the aircraft itself is doing, what the background is storing or shedding, and how the atmosphere is behaving between sensor and target. Electromagnetic presence depends not only on the platform's payload or control link but on when and how those functions are being used. Name the governing idea and the argument clarifies: signature is relational, not absolute. Explain it and the implication follows. A platform is not simply stealthy or conspicuous. It is different kinds of visible to different observers at different moments for different reasons.

That is why cold air must be understood as both a medium and a filter. It can sharpen some forms of contrast while complicating others. Visually, Arctic scenes often flatten shape and erase depth cues, especially against snow, cloud, sea ice and weak winter light. A moving aircraft may therefore evade immediate human recognition not because it has become literally invisible but because the background has become parsimonious with the cues that make objects legible. Tracking then suffers. A contact seen once may be lost again quickly against a horizon that is no longer helping the observer organise distance and motion. The target seems to disappear not only because it moved, but because the world around it stopped offering the edges and textures on which pursuit depends.

Thermally, the story is harsher because the same cold that seems like concealment can become exposure. The popular intuition is that cold air helps a platform vanish by lowering everything towards the same temperature. Sometimes it does. But contrast is never only a matter of ambient cold. It is a matter of differential heat. A drone that is relatively quiet and visually unobtrusive may still appear conspicuously in infrared if its motors, battery housings, electronics or exhaust features stand out against cold sea ice or snow that has become a remarkably austere background. The important point is not that infrared always defeats concealment. It is that cold environments often make thermal signatures less intuitive, not less relevant. The platform that sounds discreet to the ear and looks minor to the eye may be announcing itself in another register altogether.

A useful historical example lies in Cold War surveillance over northern waters and airspace, where the problem was never only finding movement but correctly distinguishing it against cold, cluttered and often deceptive backgrounds. The relevance here is not to borrow strategic theatre wholesale, but to notice a durable lesson: detection in harsh northern conditions has always been a multi-domain question because no single sensing mode remains sovereign for long. A second example can be taken from modern remote sensing and maritime search practice, where observers repeatedly discover that a contact's detectability changes dramatically with background rather than with the object alone. A vessel, aircraft or drone can become harder or easier to classify not because it changed, but because the sea state, ice field, cloud base or thermal context around it did. The environment edits the target.

That editing is especially important near ice edges, broken sea and cluttered littoral zones. These are places where the background itself becomes unstable in visual, thermal and radar terms. A platform passing against uniform cloud may be visually elusive; the same platform crossing a patterned ice boundary may become easier to track because the eye recovers relative motion from a better textured field. Thermally, an aircraft against cold sea ice may separate strongly, while the same aircraft against mixed water and fractured floe can become harder to classify because the background is now more varied and alive. For the observer, the challenge is not only to detect a signal but to decide whether it belongs to the target, the environment or some entanglement of both. That is why identification is often harder than raw detection. Seeing something is cheap compared with knowing what sort of something it is.

Auroral effects deserve care here because they tempt both exaggeration and neglect. The aurora does not turn the Arctic into a magical electromagnetic trap, nor is it irrelevant to sensing conditions. At high latitude, ionospheric behaviour can complicate some communications and radio-dependent functions in ways that alter the reliability of detection, tracking or data return without providing a neat signature drama visible to the naked eye. The lesson is not to mystify the sky. It is to remember that the Arctic atmosphere is active in layers humans do not directly perceive. A system relying on apparently clean electronic or radio assumptions may discover that the medium has become more capricious than the interface admits. Detection and identification are then strained not only by the target's signature but by the behaviour of the path between sensor and target.

This is where acoustic signatures become interesting precisely because they are so often discussed lazily. A platform can be acoustically quiet in one operational sense and still not be acoustically inconsequential. Wind direction, surface reflection, sheltering terrain, ice characteristics and observer position all alter what quiet means. Small aircraft or multicopters may disappear quickly into ambient noise under some conditions and become surprisingly localised under others, especially where the environment channels or reflects sound in ways designers did not prioritise. Acoustic detection also interacts with expectation. In low-visibility and low-contrast conditions, a faint sound may prime the observer to search differently and to notice a target the eye would otherwise have missed. Signature domains do not stand apart politely. They recruit one another.

Still, a fair limitation should be admitted. Multi-domain visibility does not mean every platform is easily detectable from every angle in every condition, nor that the Arctic always favours the defender or the observer. Plenty of contacts remain fleeting, ambiguous or misclassified. Some aircraft genuinely reduce one or more signature channels effectively enough to complicate detection materially. Background and weather sometimes help the drone rather than the sensor. It would be simplistic to suggest that cold air merely exposes everything more clearly to anyone looking.

But that limitation only sharpens the main point. The right question is not whether a platform is visible in some absolute sense. It is how its visibility migrates across domains as conditions change. A mature organisation therefore resists single-adjective thinking. Quiet to whom? Hot against what? Obvious at which phase of flight? Emitting on what schedule? Tracked by which sensor through which atmospheric path? The reconciliation is intellectually demanding but operationally useful: a platform's signature should be treated as a changing profile of liabilities rather than as a stable essence. That is the only serious way to discuss detection in a region where the background itself is half the battle.

This has direct consequences for tracking as distinct from first sighting. Initial detection may come from one domain, but maintaining track often requires another. A faint acoustic cue might trigger a visual search. A visual contact might be reacquired thermally. An electromagnetic hint might help narrow space enough for the eye or radar to take over. The stronger the environment's clutter and the more variable the background, the more important these handoffs become. A system or operator who expects one sensor mode to remain authoritative throughout is likely to lose continuity exactly when continuity matters most. Detection is an event. Tracking is a discipline of moving between imperfect truths without pretending they are the same truth.

The quiet motif here is the coat hook by a door. A coat on a hook can be unremarkable in one light, suspicious in another, and obvious once it casts a shadow or catches a glint. The object has not changed nearly as much as the conditions of seeing it. Arctic detection works in much the same way. The target's presence matters, but so do angle, background, thermal context, signal path and what the observer is primed to notice. A platform that looks modest in one doorway may declare itself in the next.

The deeper point is that signatures are part of dependence. Every time a programme congratulates itself for reducing one form of observability, it risks becoming blind to what has merely been displaced into another domain. An aircraft engineered to be less acoustically prominent may impose heavier thermal or electromagnetic burdens. A platform that blends visually at range may become easier to detect during approach, loiter or data transmission. A system that appears discreet in open sky may stand out brutally against cold sea ice. Once that is understood, the question ceases to be whether the aircraft is hidden. The question becomes which domain is currently carrying the visibility tax and whether the organisation has been honest enough to account for it.

The unresolved test, then, is not whether a drone can be made quieter, smaller or harder to see in some single flattering frame. The sharper question is whether its operators understand how cold air, cluttered backgrounds and a restless upper atmosphere keep redistributing the terms of detection, so that the platform may be obscure in one domain only to become unmistakable in the next - and whether they can still recognise that shift before the environment recognises it for them.

Resilience through redundancy and doctrine

In October 1962, during the Cuban Missile Crisis, the systems that mattered most were not the ones that promised perfection. They were the ones that retained coherence under pressure: multiple channels of communication, overlapping lines of authority, fallback procedures, and organisations disciplined enough to keep acting when certainty had already thinned. That lesson has aged well. Under stress, survival belongs less to the strongest single element than to the arrangement that can afford not to trust any one element too much. One line is fragile.

That is the right place to begin any serious account of resilience in Arctic autonomous systems. People often hear the word redundancy and imagine a shopping exercise: more sensors, more radios, more vehicles, more batteries, more backups. Those things can matter. Yet the Arctic is a ruthless critic of naive multiplication. Additional hardware can introduce fresh maintenance burden, new failure interactions, extra training load and false confidence if all the added parts still depend on the same assumptions about weather, navigation, communications or operator interpretation. A duplicate weakness is still weakness. Resilience, properly understood, is not abundance by itself. It is the disciplined refusal to let one damaged dependency govern the whole mission.

This is why diversity matters as much as redundancy, and why doctrine matters as much as both. Show the problem in practical terms and the logic becomes plain. A drone loses confidence in one navigation source during poor visibility and contested communications. If the backup source depends on the same propagation conditions, the same map assumptions or the same interpretation pipeline, then the apparent spare is not a real spare at all. Likewise, if a second operator console, second aircraft or second radio link can only be used by procedures that no one has rehearsed under stress, the organisation has purchased a reserve it cannot honestly spend. Name the principle and the argument sharpens: resilience is independence plus trained use. Explain it, and the implication follows. A fleet or system becomes harder to break only when its alternatives fail differently and its people know when to trust them.

The Arctic makes this sterner because the environment has already been degrading systems before any adversary acts. Cold narrows battery behaviour. Wind compresses recovery margins. Low contrast weakens visual cross-checks. Sparse infrastructure limits repair and diversion. Weak or variable communications make centralised rescue less reliable. Under these conditions, resilience is not the luxury of a well-funded programme. It is the minimum price of operational honesty. A platform without real fallback is not simply vulnerable to hostile action. It is already living on borrowed certainty supplied by good weather, cooperative signals and uneventful handling. The first serious pressure merely presents the bill.

That is why doctrine deserves equal standing with hardware. A system may carry multiple sensing modes, several communications paths and a range of safe states, yet still fail badly if the organisation has never decided what those alternatives are *for*. Doctrine is where technical possibility becomes authorised behaviour. It tells crews what to simplify first, what to distrust first, what to preserve when priorities collide, and when continuing the mission stops being a mark of seriousness and starts becoming a refusal to admit that the conditions have changed. Without that structure, redundancy can actually increase confusion. More options mean more branches, more hopes, more opportunities to interpret the situation differently from one another, and more ways to postpone the sober decision that margin is being consumed faster than recovery can justify.

A useful historical example lies in commercial aviation's treatment of abnormal procedures and backup systems. Modern aircraft are full of redundancy, but the industry learned long ago that duplicated components alone do not create safety. What matters is how failures are prioritised, how crews are trained to recognise degraded states, and how the whole organisation decides which failures may be contained and which must force simplification. The relevance to Arctic autonomous systems is direct. A drone or remote vehicle does not become resilient because it has several sensors or several ways to communicate. It becomes resilient when those alternatives are tied to clear behavioural transitions that have been tested and rehearsed under conditions resembling the ones in which they will actually be needed.

A second example comes from NASA's remote and planetary robotics, where resilience has often depended less on raw hardware toughness than on a blend of diversity, procedural discipline and cautious degradation. Publicly known missions have succeeded because they were designed not only to keep functioning when something went wrong, but to keep *making sense* when something went wrong. The machine narrows its ambitions, preserves its state, and remains interpretable to the humans responsible for it. That is the deeper lesson worth carrying north. Resilience is not the ability to do everything after damage. It is the ability to remain useful and legible while doing less.

Still, a fair limitation should be admitted. Redundancy and diversity are not free goods. More hardware adds weight, power demand, maintenance complexity and potential disagreement among subsystems. More doctrine can slow action or encourage ritual rather than judgement. More training can consume scarce time and still fail to reproduce the exact combinations of weather, denial and human strain that a real Arctic event will impose. There is a point at which resilience architecture becomes clutter rather than strength. It would be simplistic to suggest that every additional backup or every added layer of procedure automatically improves survivability.

But that limitation only clarifies the real standard. The serious question is not how many backups exist. It is whether the programme has created *meaningful* alternatives. A meaningful alternative does not share the same failure path, does not demand the same fragile assumptions, and does not arrive so late or in so confusing a form that operators cannot use it honestly. The same standard applies to doctrine. A meaningful doctrine is not a thick manual of admirable intentions. It is a practical, remembered, tested set of behavioural reductions and priorities that people can carry into ugly situations without pretending the ugly situation is still normal. The reconciliation is therefore stern and useful: resilience should be judged by the quality of degraded operation, not by the splendour of nominal capability.

This is why disciplined modes of operation are so important under pressure. Systems in contested and environmentally harsh conditions should not merely struggle onwards until forced down to their last branch. They should know how to become smaller versions of themselves. That may mean narrowing mission scope, refusing certain manoeuvres, simplifying sensor use, reducing reliance on weak links, preserving logs over imagery volume, or prioritising safe recovery over elegant persistence. None of this sounds heroic. That is precisely the point. A resilient system is often one that disappoints ambition early enough to preserve truth later. Programmes that cannot bear such disappointment usually discover, too late, that they were mistaking uninterrupted activity for robustness.

The human dimension makes this unavoidable. Operators, maintainers and commanders are part of the resilience architecture whether or not the programme says so out loud. If training has taught them to admire completion more than disciplined retreat, they will misuse redundancy by spending one backup after another in the service of a mission already made dishonest by changing conditions. If doctrine has not taught them which signs matter most when denial and degradation combine, they will treat every option as still open until most of them have silently expired. If organisational culture punishes uneventful aborts or conservative degradation, then the hardware will be forced to carry a burden no hardware can carry alone. Resilience is therefore social before it becomes mechanical.

The quiet motif here is the spare key. A spare key is not useful merely because it exists in principle. It must be cut differently enough to matter, stored where it can be reached when the lock fails, and trusted by the person holding it to open the right door at the right time. Otherwise it is a comforting object with no real operational value. Redundancy in Arctic systems often behaves in much the same way. Spare sensors, spare links, spare vehicles and spare modes are meaningful only if they can be reached under stress and if their use has already been made intelligible before the crisis arrives.

This also changes how one should judge claims of resilience in public or institutional language. The strongest question is not whether the system has backups. It is what sort of behaviour those backups enable when the machine is under simultaneous environmental and adversarial pressure. Does the system remain governable? Do the alternatives fail differently? Can the organisation explain, in advance, what will be sacrificed and what will be preserved? Can operators practise degraded modes without treating them as embarrassing imitations of the real mission? Does the whole structure keep the aircraft or vehicle inside a future the organisation can still understand, recover and defend? These are not decorative questions. They separate resilience from inventory.

The unresolved test comes when one dependency has already weakened, another is beginning to wobble, and the mission still appears just possible enough to tempt continuation. At that moment, does the programme possess genuine resilience - independent alternatives, disciplined doctrine and trained restraint - or merely a collection of extra parts and hopeful words that will fail in the same direction once the pressure becomes real?

Escalation control and risk management

On 27 October 1962, during the Cuban Missile Crisis, the United States destroyer USS *Beale* and other ships dropped signalling depth charges near the Soviet submarine B-59. The officers on both sides were operating inside an atmosphere thick with uncertainty, technical strain and partial understanding. The most important question was not who possessed the stronger platform. It was who had already decided, before the next shock arrived, what counted as acceptable pressure, acceptable loss and acceptable response. The margin for improvisation was smaller than anyone wanted to admit. Stress reveals doctrine.

That is the right place to begin any serious account of escalation control for autonomous systems in the Arctic. People often hear the word risk and imagine probabilities, hazard tables or broad managerial language about trade-offs. Under contested conditions, risk is much more concrete. It is the lived question of how much loss, uncertainty and machine initiative an organisation is prepared to tolerate before it changes the mission, narrows authority or accepts that recovery is no longer worth the price it may demand. If those limits are not explicit before the system is stressed, they will be invented under pressure by whoever happens to be holding the least incomplete picture at the worst possible moment. That is not management. It is drift.

This matters because autonomy changes where escalation first becomes visible. In a crewed system, danger often presents itself as rising human exposure. In an autonomous or remotely supervised system, danger can appear first as a machine still moving but becoming less governable, less legible or more expensive to recover honestly. Show the practical sequence and the point sharpens. A platform enters degraded conditions, suffers uncertain interference, begins losing confidence in one or more dependencies, and still seems physically capable of continuing. The temptation is immediate: press on, because the airframe remains aloft and the mission may still be possible. Name the real issue and the argument clarifies: control limits under stress. Explain it and the implication follows. The serious question is not whether the system can continue, but whether continued operation still fits the organisation's declared tolerance for ambiguity, loss and unintended consequence.

This is why acceptable loss deserves to be treated as a prior policy choice rather than as an emotional reaction after something goes wrong. Organisations often avoid this question because it feels harsh or politically awkward. Yet the avoidance is itself dangerous. If a programme has not decided what kinds of loss are acceptable under which conditions, then every difficult event becomes morally improvised. Is the loss of an airframe tolerable if data custody is preserved? Is an emergency landing preferable to continued exposure to deception? Should a compromised machine be recovered at greater risk, or should it be abandoned if recovery would enlarge the problem? These are not matters that should first be discovered in the weather, under interference, with the aircraft already narrowing its options. They belong to governance before they belong to drama.

A useful historical example lies in maritime nuclear command and control during the Cold War, where the most responsible systems were not those promising unrestricted response, but those defining thresholds, authorities and inhibitions clearly enough to prevent technical stress from becoming uncontrolled political movement. The relevance here is structural, not sensational. Autonomous Arctic operations face a humbler but recognisably similar question: what has the organisation already decided the machine may and may not do when information becomes unreliable and the environment begins compressing time? The lesson is not that every drone mission is strategically equivalent to crisis diplomacy. It is that escalation is often a problem of predeclared limits rather than of raw technical capability.

A second example can be found in modern civil aviation's treatment of abnormal procedures and go/no-go criteria. Mature aviation culture does not treat every aircraft that remains physically flyable as equally fit to continue the mission originally imagined. It distinguishes among states of confidence, authority and recoverability. That distinction is deeply relevant to autonomous systems. A drone that remains airborne under pressure is not thereby entitled to continue gathering data, reattempting recovery or extending the problem space around itself. A resilient organisation accepts that surviving aloft and remaining operationally honest are different achievements. Escalation control begins when those achievements are no longer confused.

This is also why autonomy behaviour under threat must be bounded in advance. A platform under contested conditions may need to simplify, retreat, hold, divert, reduce emissions, preserve records or relinquish parts of the mission. Those are serious actions even when described at a conceptual level. What matters here is not the specific behaviour in a tactical sense, but the existence of a disciplined hierarchy of machine responses that reflect organisational priorities. If no such hierarchy exists, autonomy becomes vulnerable to two opposite errors. It may remain too passive, persisting with a brittle plan because it lacks authority to narrow its own ambition. Or it may become too adventurous, exercising forms of initiative that the organisation later disowns because no one had settled, in advance, what counted as legitimate machine discretion under threat. Both failures begin in the same place: undeclared limits.

Still, a fair limitation must be admitted. Limits set too rigidly can waste capability, induce premature abandonment and reduce the flexibility that sometimes distinguishes a resilient programme from a timid one. Harsh environments and contested conditions do not always produce the same kind of pressure, and an organisation that writes its thresholds too mechanically may discover that it has taught the machine and its operators to stop when more nuanced judgement was still possible. This is a genuine criticism. Escalation control can become bureaucratic if it is designed as an abstract compliance exercise rather than as a living expression of operational reality.

But that limitation only clarifies the deeper standard. The solution is not vagueness. The solution is explicitness joined to review. Control limits should be clear enough to guide action under stress and revisable enough to remain faithful to field evidence rather than ceremonial caution. That is a sterner demand than it sounds. It requires organisations to say, in plain language, what they are protecting when systems come under pressure. Is the priority preserving the airframe, protecting data credibility, preventing wider exposure, retaining human authority, or avoiding recovery attempts that create secondary risks? Once that order is declared, risk language becomes sharper and more honest. Without it, people reach for the word risk while meaning entirely different things.

This is why recovery attempts deserve special scrutiny in contested conditions. Recovery is often treated as the natural moral endpoint of any sortie: if the aircraft still exists, bring it back. Yet under threat, recovery can become its own escalation mechanism. One more approach, one more loiter, one more search pattern, one more relay, one more attempt to preserve a platform that has already ceased to be trustworthy can enlarge the operational and evidential damage rather than reduce it. A serious organisation therefore asks not only whether recovery is possible, but whether it remains proportionate to the value and condition of what is being recovered. That proportionality is rarely flattering. It often demands that people accept a smaller, duller outcome than their instincts or sunk costs would prefer.

The quiet motif here is the fuse. A fuse is not a sign of weakness in an electrical system. It is a declaration that uninterrupted function matters less than preventing a larger failure from propagating. Escalation control should serve the same purpose in autonomous operations. It is not there to make the system look bold. It is there to trip before stress, loss and ambiguity begin travelling farther than the organisation can still explain or govern. A machine without such fuses may appear more capable for a while. It is often merely better at hiding that it has already exceeded the human terms of its permission.

This has direct implications for how risk should be discussed by thoughtful professionals. The useful language is not melodramatic and not evasive. It asks which losses are tolerable, which uncertainties are tolerable, which machine behaviours remain authorised under degraded confidence, and which attempts at preservation or recovery become self-defeating once threat and environment have combined. It distinguishes survivability from usefulness, persistence from honesty, and capability from licence. That distinction is what prevents risk talk from sliding into either empty caution or empty bravado. A programme that cannot articulate those distinctions clearly will eventually substitute mood for doctrine.

The deeper point is that escalation control is not really about the adversary alone. It is about how an organisation disciplines itself when the adversary, the environment and the machine's own dependencies begin pushing at the same time. If acceptable loss is undefined, autonomy behaviour vaguely bounded and recovery attempts guided mainly by hope, then the system will improvise its ethical limits in flight. That is too late. The most serious programmes understand that control limits are part of system design, not merely part of command culture. They are how institutions keep their own ambitions from becoming the most dangerous thing in the battlespace.

The unresolved question, then, is not whether a particular platform can survive one more attempt, one more minute or one more degraded mission segment under pressure. The harder question is whether the organisation has already decided what it is unwilling to spend - in airframes, in authority, in trust, in recoverability and in explanation - before the system is stressed enough to make improvisation feel courageous and discipline feel like surrender.

Chapter 13

Testing Regimes and Qualification Standards

What should the words Arctic-ready actually buy you: confidence, or merely relief from having to ask harder questions? The phrase sounds practical until one tries to pin it down. A platform survives a cold morning, completes a trial, returns with impressive footage, and someone says the system is ready. Ready for what, exactly? For one benign run in northern weather, or for repeated use after cold soak, gloved maintenance, blowing snow, uncertain recovery, darkness, moisture and the slow accumulation of small indignities that turn a promising machine into a seasonal disappointment?

That is the right discomfort to begin with, because readiness language often arrives long before readiness evidence does. Organisations like the phrase Arctic-ready for understandable reasons. It compresses engineering effort, procurement aspiration and operational hope into something that can travel cleanly across briefings and budgets. But the Arctic is severe about adjectives. It does not care what a system was called in the design review, only what the system can still do after cold has altered its materials, fine snow has found its way into neglected seams, wind has distorted the expected recovery geometry, and a maintainer in gloves and poor light has had to prepare it for another sortie without pretending away uncertainty.

That is why qualification in this theatre is not simply a technical threshold. It is a moral one. A system declared ready without evidence matched to the conditions of use does not merely risk embarrassment. It teaches operators, maintainers and commanders to trust a claim whose meaning has never been made fully explicit. The danger lies not only in dramatic failure. It lies in the quieter culture that grows around under-tested systems: repeated anomaly seen as personality rather than evidence, rough-weather abort treated as bad luck rather than design truth, field workarounds mistaken for proof of resilience, and isolated successful trials inflated into institutional confidence. The label begins to do the work the data never did.

The central tension lies there. Testing regimes want control, repeatability and legible results. Arctic operations produce interaction, surprise and compound stress. Chambers, tunnels and accelerated ageing programmes are indispensable because they allow engineers to isolate variables and learn precisely. Field trials are indispensable because isolation is not how the environment actually behaves. Standards want clarity. Governance wants milestones. Engineers want evidence. Operators want systems that remain honest under pressure. The difficult truth is that these desires align only if the testing philosophy itself admits that confirmation is not enough. The purpose of qualification is not to prove that the machine can succeed under selected conditions. It is to discover whether the machine still deserves trust when the conditions stop being tidy.

That matters because cold-weather failure is so often combinatory rather than singular. A component may tolerate temperature in a chamber and still become doubtful when vibration, moisture and contamination join it in the same sequence the field will actually impose. A sensor may pass a narrow environmental test while remaining poorly matched to the snow ingress, handling stress or thermal transitions that accompany real deployment. A maintenance task may look reasonable on a bench and become absurd in darkness with gloves. If testing does not force those combinations into view, the institution will continue speaking as though readiness were a property of hardware rather than a relationship between hardware, procedure, environment and time. That misunderstanding is expensive.

The stakes therefore extend beyond engineering. Procurement timelines and reporting structures shape what gets called evidence, what gets deferred into later spirals, and what never receives serious scrutiny because it threatens programme smoothness more than it threatens the machine in the present. A regime organised around passing gates will naturally favour tests that are clean to schedule, easy to narrate and flattering to the design. A regime organised around learning will tolerate more discomfort, more awkward data and more results that complicate the story of progress. The Arctic does not reward the first culture for long. It simply waits until the avoided questions return in the form of field failures, shortened confidence intervals and expensive post hoc explanations.

That is why readiness debates should be read institutionally as well as technically. When someone claims a platform is Arctic-ready, the reader should hear several quieter questions at once. Ready according to which test logic? Ready after which combinations of cold, wind, moisture, icing, handling and darkness? Ready for a single demonstration or for repeatable sustainment across a season? Ready in the hands of the engineers who built it, or ready in the hands of ordinary operators and maintainers working under the shelter, tool and data conditions the theatre will actually provide? These are not cynical questions. They are the minimum terms of seriousness.

The path ahead follows those terms from definition into consequence. It begins with the stubborn question of what Arctic-ready should mean when stripped of romance and tied to observable proof. It then moves into the laboratory and its blind spots, where controlled testing earns its authority and also meets its limits once stressors start interacting. From there the argument turns to field trials and telemetry, where the quality of instrumentation decides whether disappointment becomes learning or merely a more expensive anecdote, before arriving at standards, procurement and reporting structures, where institutional incentives decide how much inconvenient truth a programme is willing to pay to know.

A useful way to hold these questions together is to think of a winter coat bought by catalogue description alone. On paper it may look ideal: insulated, weather-resistant, durable, designed for the cold. The real judgement arrives outside, in wind, wet, movement and repeated use, and later still when the zip stiffens, the cuffs soak through and the pockets become impossible in gloves. A machine tested for Arctic work demands the same kind of honesty. Specification is the beginning. Repeated contact with conditions is the verdict.

The harder question, then, is not whether a system can survive some cold trials and emerge with a persuasive dossier. The sharper question is whether its evidence has been gathered in a way that lets operators, maintainers and institutions trust the claim once the environment begins combining cold, moisture, wind, handling and governance pressure into the very situations the qualification regime found hardest to reproduce.

What "Arctic-ready" should mean

In April 1912, when Roald Amundsen's expedition machinery and equipment were judged against Antarctic conditions, the decisive standard was never whether a thing worked once in the cold. It was whether it remained dependable after cold had ceased to be an event and become the whole terms of life. Sledges, clothing, stoves and dogs were not praised for surviving an ordeal in isolation. They were valued because they could be used, maintained and trusted repeatedly in an environment that did not forgive sentiment. The weather kept asking the same question.

That is the right place to begin any serious definition of Arctic readiness for autonomous systems. Too often the phrase Arctic-ready is awarded after a demonstration, a winter exercise, a successful sortie or a vendor's account of northern use. None of that is worthless. None of it is enough. A platform that launches and returns in cold weather once may have proven only that one crew, one configuration and one set of margins happened to align on one day. The Arctic is harsher than that. It asks whether the machine can be soaked by cold long enough for materials, batteries, seals and displays to become the versions of themselves that operators will actually inherit. It asks whether the aircraft can face icing as an operational condition rather than as a laboratory curiosity. It asks whether wind tolerance means recovery under ugly air, not merely transit through broad forecast limits. It asks whether a human being in gloves and darkness can still maintain and prepare the system without turning every routine task into a gamble. Ready is a repeated condition.

This is why readiness should be defined as a chain of claims tied to repeatable tests rather than as a flattering mood around a platform. Show the ordinary temptation and the problem becomes obvious. A system performs well after being carried from warm storage into cold air. It flies a mission, returns, and the event enters organisational memory as proof that the aircraft has passed some unwritten northern threshold. Name the weakness in that reasoning and the argument sharpens immediately: anecdotal qualification. Explain it and the implication follows. What was demonstrated may have been only brief exposure, only limited handling, only partial environmental truth. If the machine was not cold-soaked through its full structure, if the maintainers were not working in realistic darkness or bulked clothing, if icing was absent, if wind remained cooperative, then the claim of readiness outruns the evidence. The phrase has been allowed to do more work than the test.

Temperature soak belongs at the centre of this because cold is not merely ambient. It is cumulative. Components cool at different rates, housings lag behind exposed surfaces, batteries, lubricants and displays do not all arrive at the same state at the same time, and recovery from cold can be just as consequential as exposure to it. A platform carried briefly into sub-zero air may appear competent while much of its mass still reflects the kindness of previous shelter. A truly Arctic-ready machine should be judged after the easy thermal memory is gone. That means the system has to be treated not as a device encountering cold, but as a cold object asked to perform. The difference matters because many failures emerge only after cold has had time to rewrite the machine's internal negotiations: stiffness, shrinkage, lag, power penalties, condensation paths, sensor drift and altered operator interaction. Cold must become the baseline, not the surprise.

Icing exposure deserves equal seriousness for the same reason. Ice is often discussed as though it were a binary disaster: either the aircraft ices badly and fails, or it does not. Real operations are less theatrical and more difficult. Light accretion, intermittent contamination, partial sensor obscuration and the altered feel of control or drag can all degrade the system before any single dramatic threshold is crossed. That is why readiness claims must not rely on a simplistic declaration that the platform has "seen icing". The meaningful question is whether the platform has been tested in the kinds of icing exposure that affect the whole chain of use: flight behaviour, sensing, recovery, inspection and subsequent reservice. A machine that flies through some cold moisture once but cannot then be cleaned, checked and relaunched honestly is not ready. It has merely been lucky in one segment of the cycle.

Wind tolerance is often treated with similar laziness. A manufacturer or operator may cite the fact that the aircraft remained stable or completed a mission in strong wind. Yet the Arctic punishes this kind of broad statement because wind matters differently across phases of flight. A platform may travel capably in open air while becoming unreliable or operationally dishonest in the disturbed, gusting, low-margin air of launch and recovery. Readiness should therefore never be inferred from cruise behaviour alone. The real standard is whether the system can still recover with repeatable authority, clear abort logic and credible margins once the wind becomes local, variable and entangled with terrain, deck motion or poor visibility. Anything less is not a weather limit. It is a weather hope.

A useful historical example lies in aviation certification culture more broadly, where serious airworthiness has long depended on repeatable, documented evidence rather than on pilot anecdotes, however sincere. The relevance here is structural. No mature aviation regime would accept "we flew it once in difficult weather" as the equivalent of demonstrated capability across the envelope that actually matters. Arctic systems deserve the same sternness. A second example can be drawn from polar field science, where instruments and vehicles often perform beautifully when first exposed and then reveal their true character only after repeated thermal cycling, awkward handling, storage transitions and the cumulative small insults of real deployment. The lesson worth borrowing is plain: harsh-environment proof should be earned through recurrence, not through memorable exceptions.

Still, a fair limitation should be admitted. No test regime can reproduce the entire Arctic, and no definition of readiness can abolish uncertainty once a platform leaves the trial plan and enters weather, darkness, human fatigue and seasonally shifting surfaces. Overclaiming is dangerous. So is the fantasy that readiness can be certified into perfection. A sensible qualification standard should not promise immunity from harsh conditions. It should promise that the organisation knows what has been tested, what remains uncertain, and how the machine behaves when pushed near the limits that actually matter. The purpose of the phrase Arctic-ready is not to end argument. It is to narrow argument to evidence the institution can defend.

That is why maintainability under gloves and darkness belongs inside the meaning of readiness rather than at the edge of it. It is tempting to treat operator interaction as a separate human-factors issue and readiness as a property of the vehicle itself. The Arctic does not permit that separation. A machine requiring bare-handed dexterity, clean bench conditions or bright, spacious work areas for routine preparation and inspection is not ready for the theatre merely because the aircraft in flight behaves well. Readiness includes the truth of handling. Can access panels be opened, inspected and closed with realistic gloves? Can connectors, seals, batteries and critical interfaces be manipulated in the lighting conditions the mission will actually impose? Can the crew perform ordinary checks without warming the system into a different thermal state than the one in which it must fly? These are not side questions. They decide whether the system can move from one mission to the next without accumulating hidden doubt.

This is where evidence must become matched evidence. A platform may pass cold tests in a chamber, icing exposures in a controlled setup, wind trials in useful conditions and maintainability checks in a workshop. Each result may be individually valid. What matters is whether together they speak to the same machine, in the same operational reality, across the same repeated cycle of use. The Arctic is a theatre of interactions, not isolated adjectives. A credible readiness claim therefore ties temperature soak to recovery, icing to inspection, wind to abort logic, gloved maintenance to night conditions, and all of these to a system that is asked to repeat rather than merely to impress. The more fragmented the evidence, the more suspicious the adjective should become.

The quiet motif here is the door latch. On a warm day, with bare fingers and good light, it may click shut with satisfying certainty. In darkness, through thick gloves, after thermal shrinkage and repeated cold cycling, the same latch may require a different hand, a second look or a degree of trust it no longer deserves. Arctic readiness is full of such latches. Small interfaces that were invisible in ordinary testing become the true examiners of the whole claim. A serious qualification culture listens to them.

This has consequences for how a thoughtful reader should interpret procurement language and programme briefings. The right response to an Arctic-ready label is not awe and not cynicism. It is a series of disciplined questions. Ready after what sort of soak? Ready under what icing burden? Ready for what kind of wind in what phase of flight? Ready for whose hands, in what gloves, under what light, after what maintenance cycle? Ready once, or ready again? Such questions do not obstruct progress. They rescue the word ready from turning into a ceremonial compliment. The phrase gains value only when it can survive those questions without retreating into anecdote.

The deeper point is that readiness is not a trait hiding inside the aircraft, waiting to be discovered like a metal property. It is an earned relationship among machine, environment, operator and institution. It depends on whether the tests resemble the dependence they are claiming to support. An organisation that bases readiness on stories will eventually learn by disappointment. An organisation that bases it on repeatable, condition-matched evidence may still be surprised by the Arctic, but it will at least be surprised honestly.

The unresolved question, then, is not whether a given platform has shown that it *can* work in the cold. Many can. The sharper question is whether the evidence proves that the whole system - aircraft, recovery, maintenance, human handling and repeated service - remains trustworthy after the Arctic has had time to make everything colder, harder, slower and less forgiving than the original success story allowed.

Environmental testing and its blind spots

In 1986, after the loss of the Space Shuttle *Challenger*, one of the most enduring lessons was not simply that hardware can fail in the cold. It was that a test regime can be technically serious and still leave the most dangerous interaction insufficiently understood. The individual pieces of the story had been visible. The way they would combine, under real pressure and real timing, had not been honoured with equal force. The test was not worthless. It was incomplete.

That is the right place to begin any discussion of environmental testing for Arctic systems. Chambers, wind tunnels and accelerated ageing programmes are indispensable precisely because they let engineers isolate variables, repeat conditions and build arguments stronger than anecdote. Yet those strengths can seduce. A system that passes a cold chamber test, a vibration run and a moisture exposure may still fail in the field because the Arctic does not deliver its insults one by one and with polite intervals between them. It layers them. Cold changes material behaviour, vibration loosens what was only just acceptable, moisture migrates, fine snow finds edges and seams, and a component that seemed individually qualified enters service as part of a much rougher conversation. The machine survives the test and loses the winter.

This is why blind spots in environmental qualification are rarely caused by laziness alone. More often they arise from the very logic that makes laboratory testing respectable. Show one variable clearly, control the rest, learn what the system does, then move to the next question. That method is powerful. It is also partial. Name the real difficulty and the argument sharpens: interaction blindness. Explain it and the implication follows. The more a programme relies on tests that isolate stressors cleanly, the greater the risk that it will overstate confidence in systems whose real field behaviour depends on how stressors meet, overlap and accumulate in the same hours, on the same surfaces, through the same tolerances and seals. The Arctic is not a neat sequence. It is a stack.

Cold chambers reveal much that no serious programme should ignore. They show whether batteries sag, displays slow, lubricants stiffen, plastics embrittle, seals shrink and software assumptions about warm hardware begin to look foolish. Wind tunnels likewise matter because they make flow visible, expose load behaviour and help separate aerodynamic fact from impression. Accelerated ageing helps organisations ask harder questions about materials and repeated cycles than a short field season can answer by itself. All of that is real value. The mistake begins when people allow the reality of those values to harden into the belief that environmental truth has therefore been captured in full. A chamber is a question asked cleanly. The field answers messily.

A useful historical analogue lies in aviation certification more broadly, where components and systems have long been tested under rigorous environmental standards and still later revealed awkward interactions in service. The point is not that formal testing fails as a concept. The point is that formal testing succeeds at what it is designed to ask, and nothing more. A second example comes from remote and polar instrumentation, where publicly described equipment often passes low-temperature trials and later proves troublesome once snow ingress, transport vibration, repeated handling and condensation cycles join the story. The pattern is instructive because it is so ordinary. The field rarely announces a brand-new law of physics. It merely combines familiar ones in a timing the qualification regime had not forced into the same frame.

This is why combined-stressor testing matters so much. A sensor that passes cold tests in dry conditions may still fail once cold is paired with fine snow ingress, not because either factor is exotic in isolation, but because together they alter friction, visibility, sealing and the meaning of tolerance at the same time. A connector that survives vibration in warmth may become intermittent when vibration is followed by thermal cycling and then by moisture migration during storage. A control surface or gimbal may behave acceptably in cold still air yet degrade once gust loading, icing residue and post-flight handling are added to the sequence. The important phrase here is sequence as much as combination. Some failures are not merely co-produced by multiple stressors. They are produced by the order in which the stressors arrive.

That point is easy to miss because organisations often think in categories while the Arctic behaves in chains. A chamber belongs to temperature. A tunnel belongs to flow. An ingress trial belongs to contamination. An ageing test belongs to time. The aircraft, unfortunately, belongs to all of them at once. It launches after storage, flies after soak, lands through gusts, is cleaned in awkward shelter, repacked while thermally uneven, and emerges next day carrying the history of each step in every seal, bearing, battery, window and fastening interface. A serious qualification regime therefore asks not only "Can this component survive cold?" but "What else will cold be doing to the conditions under which this component next meets moisture, vibration, handling or delayed maintenance?" The answer is often where blind spots stop being abstract.

Still, a fair limitation deserves emphasis. No testing organisation can reproduce the full Arctic with perfect fidelity, and no qualification campaign can combine every stressor in every plausible sequence without becoming impossible to finish or impossible to interpret. At some point the programme must choose what combinations matter most and accept that uncertainty remains. It would be foolish to imply that the remedy is simply to test everything with everything else forever. Exhaustion is not rigour. Confused data is not wisdom.

But that limitation only clarifies the real standard. The question is not whether every combination can be tested. The question is whether the combinations most likely to produce operationally serious misunderstanding have been chosen honestly. A system intended for snowbound, high-wind, repeatedly handled service should not be qualified mainly through elegant single-variable success. It should be challenged where its dependencies intersect: cold with ingress, vibration with thermal cycling, moisture with storage delay, darkness and gloves with maintenance tasks that looked trivial on the bench. The reconciliation is demanding but practical. Test the interactions most likely to turn nominally acceptable parts into field-liability behaviour, then remain explicit about what has still not been learned.

This is also why telemetry and instrumentation matter inside the laboratory, not only later in field trials. A combined-stressor test that merely ends in pass or fail may flatter the programme with decisive language while revealing very little about mechanism. Good instrumentation lets the system fail in a way that teaches. It shows drift before breakdown, latency before lock-up, changing motor current before seizure, degraded clarity before full sensor uselessness, rising control effort before oscillation. Without such detail, a programme learns that something went wrong under combined stress and remains tempted to attribute the result to the most visible variable. With it, the organisation can distinguish whether the true culprit was ingress enabled by cold shrinkage, vibration that altered sealing, or moisture behaviour changed by prior thermal history. Diagnosis, not disappointment, is the point.

The quiet motif here is the window latch. In a warm workshop it clicks shut with satisfying certainty. In the field, after cold soak, light icing, repeated vibration and a little fine snow, the same latch may still close and yet no longer mean what it once meant. Much of Arctic qualification lives in such latches: parts and interfaces that continue to appear acceptable until several ordinary stressors have combined to make their old signal unreliable. Environmental testing earns its authority when it learns to distrust that first click.

This has institutional consequences as well. Programmes under schedule pressure tend to prefer tests that are legible in reports and portable in procurement language. Single-stressor success is easier to narrate, easier to compare and easier to turn into tidy readiness claims. Combined-stressor evidence is often uglier. It raises awkward questions about redesign, maintainability, inspection intervals and what exactly the words qualified or ready should mean. Yet those awkward questions are not inefficiencies in the testing regime. They are the regime finally becoming honest enough for the theatre it serves. A clean report that hides interaction risk is not a bargain. It is deferred embarrassment.

The deeper point is that blind spots are not only technical omissions. They are often governance decisions disguised as testing philosophy. What is chosen for combination, what is left separate, what counts as representative, what is instrumented richly enough to teach, and what is declared good enough for programme movement all express an institution's appetite for inconvenient truth. Arctic operations punish institutions that underprice that truth. The weather does not care which stressor was assigned to which lab team or budget line. It only asks whether the system that emerges from qualification understands the world it is claiming to enter.

The unresolved question, then, is not whether a platform can survive chambers, tunnels and accelerated ageing in isolation. Many can. The sharper question is whether the qualification regime has confronted the ways cold, vibration, moisture and contamination will actually meet inside the same vehicle, in the same sequence of use, before the field is left to discover that the blind spot was never in the environment at all, but in the elegance of the test plan.

Field trials, telemetry, and learning design

In 1970, the damaged Apollo 13 spacecraft turned telemetry into something more than a stream of engineering numbers. It became the only disciplined way of knowing what the machine still was, what it had stopped being, and which hopeful interpretations no longer deserved belief. The lesson was not that data always saves. It was that under stress, uninstrumented uncertainty quickly becomes mythology. The signal had to speak.

That is the right place to begin any serious account of field trials for Arctic autonomous systems. A field trial is often described as a proving ground, which sounds firmer than the reality. In harsh conditions, a trial is less a stage for proof than a controlled encounter with ignorance. The weather shifts, the platform ages, the operators adapt, the support equipment develops personality, and the system begins to reveal which parts of its earlier success were durable and which were merely flattering. A trial that cannot hear those revelations is not really a trial. It is a demonstration in outdoor clothing.

This is why telemetry matters so much. Not because more data is always wiser, but because the right data determines whether failure becomes explanation or embarrassment. Show the common scene and the issue is plain. A vehicle wobbles on approach, drops a sensor stream, shows a transient power anomaly, aborts a task, or lands with behaviour that "just looked wrong". If the trial has been instrumented only to support basic safety and celebratory reporting, then the post-flight conversation quickly fills with guesswork. Was it wind, icing, software timing, contamination, operator choice, thermal drift, a lagging controller, a misleading sensor or some intersection of several? Name the real challenge and the argument sharpens at once: diagnosis speed under uncertainty. Explain it and the implication follows. The faster a programme can move from "something failed" to "this is how it began, propagated and became visible", the more credible its learning will be.

That credibility is the whole point. A field trial is not valuable merely because it places a system in real weather. It becomes valuable when its instrumentation preserves enough of the machine's inner life that later judgement can distinguish symptom from mechanism. Basic position, attitude and system state are not enough if the mission is asking harder questions about autonomy, recovery, sensing and maintainability in the cold. Timing relationships matter. Mode transitions matter. Sensor confidence matters. Environmental context matters. Operator inputs matter. Storage and buffering behaviour matter when links thin or break. A trial designed around confirmation will collect the minimum needed to say the sortie happened. A trial designed around uncertainty will collect what is needed to understand why the sortie became what it did.

A useful historical example lies in aviation flight testing more broadly. Serious test cultures learned long ago that a successful flight without diagnostic richness is often less valuable than a troubled flight whose data is good enough to teach. The relevance here is direct. Arctic drone trials do not become mature by producing a string of apparently clean sorties. They become mature when the programme can explain its anomalies with enough precision that design, procedure and doctrine can be revised on evidence rather than on memory or rank. A second example can be drawn from remote scientific and polar instrument campaigns, where the real divide between an informative failure and a useless one often lies in whether the system preserved state, timing and environmental context rather than merely the final output. Equipment that dies noisily but legibly can teach more than equipment that appears stable until the record turns blank.

This is why good trial design starts from uncertainty rather than from affirmation. Organisations naturally want trials to support programme momentum. Schedules, budgets and reputations all encourage field events that can be narrated as progress. That pressure quietly shapes instrumentation choices. Systems are equipped to show performance, not vulnerability; to capture outputs, not transitions; to support safety, not post hoc humility. The result is a paradox: the more a trial is designed to reassure, the less able it becomes to explain the very surprises the field was supposed to reveal. A platform may return successfully while the organisation learns almost nothing except that, under a narrow set of conditions, luck and design remained temporarily aligned.

This is where telemetry design becomes an institutional argument, not just an engineering one. A rich telemetry set imposes cost: bandwidth pressure, onboard storage demands, more careful tagging, greater parsing burden, and the need for analysts capable of turning streams into judgement. Those burdens are real. Yet the absence of such telemetry imposes its own cost, usually later and more painfully. Diagnosis slows. Competing narratives multiply. Design changes become speculative. Lessons learned reports begin to sound confident in tone and vague in mechanism. The organisation then mistakes documentation for learning. A mature programme pays the earlier cost because it understands that a hard-earned anomaly without enough evidence is operationally expensive twice over: once in the sortie, and again in the confused future built on its misunderstanding.

Still, a fair limitation deserves emphasis. No field trial can capture everything. Storage is finite. Links fail. Sensors drop out. Trial teams cannot instrument every possible pathway without overwhelming the system or their own ability to interpret the result. There is also a genuine danger that programmes collect vast data volumes without deciding what questions they are actually trying to answer. That kind of richness can become its own form of blindness. More signals do not automatically mean more knowledge. Some trials drown in their own sincerity.

But that limitation only clarifies the real standard. The question is not whether every variable can be recorded. The question is whether the trial was instrumented around the uncertainties most likely to matter. If cold-weather recovery is the concern, then the data architecture should preserve what lets the team reconstruct changing control effort, sensor confidence, wind-related behaviour and operator response. If maintainability is under scrutiny, the trial should not only watch the flight but the state transitions before and after it. If perception in snow and glare is the issue, the record must preserve not just the image but the conditions under which that image was being trusted. The reconciliation is exacting but practical: design telemetry to answer the argument the field is most likely to start, not merely to decorate the sortie log.

This is also why learning design matters as much as the sensors themselves. A programme can instrument a vehicle well and still fail to learn if the post-trial process is built around defending prior assumptions. Learning design begins earlier, in the willingness to define what counts as an informative failure, what anomalies must be preserved rather than explained away, and which trial outcomes should trigger deeper investigation even when the sortie still "succeeded". That attitude changes the whole culture of testing. The team stops seeing deviation as embarrassment and starts seeing it as the raw material of qualification. Without that shift, telemetry becomes a compliance archive rather than an investigative tool.

The quiet motif here is the black box, not as a dramatic icon of disaster, but as a reminder that machines speak most usefully when their records survive their performance. A field trial needs something of that spirit even when nothing crashes and no headline event occurs. The programme should ask, after every difficult sortie, whether the record is strong enough to let the machine disagree with the most convenient human story about it. If not, the trial may have exercised the system without really interrogating it.

This changes how one ought to judge trial success. A clean flight is not necessarily a strong trial. A troubled flight is not necessarily a bad one. The stronger trial is the one that leaves the organisation better able to distinguish design weakness from procedural error, environmental penalty from poor assumption, repeatable pattern from isolated nuisance. That distinction is what gives lessons learned their credibility. Without it, the programme accumulates incidents and opinions rather than knowledge. With it, even disappointment can become a form of progress stern enough to deserve the name.

The unresolved question, then, is not whether the next field trial will produce a pass, a fail or a neatly mixed report. The sharper question is whether its telemetry and learning design are good enough that, when the Arctic introduces exactly the kind of awkward, compounded behaviour everyone claimed to welcome as realism, the failure will teach faster than the institution is tempted to tidy it away.

Standards, procurement, and institutional incentives

In 1986, when the Rogers Commission examined the loss of the Space Shuttle *Challenger*, one of the clearest conclusions was not merely that engineering had failed, but that institutions had taught engineering what kinds of failure were acceptable to postpone. Schedules had a voice. Reporting lines had a voice. The formal language of readiness had a voice. Hardware then spoke in the accent of governance. That is the lasting lesson.

It matters here because Arctic qualification is often discussed as though it were a technical argument conducted in clean air above procurement, budget cycles and organisational pride. It is nothing of the sort. A platform is tested according to what the institution is prepared to delay, to pay for, to measure, to admit, and to call incomplete. The most revealing question is therefore not only whether a system passed a certain environmental test or field trial, but why that test was chosen, why another was deferred, and what sort of programme reality made one body of evidence count as sufficient while another was quietly declared excessive. Engineering does not sit outside governance. It obeys it.

This becomes obvious the moment one looks at how readiness claims are assembled. Show the ordinary sequence and the point sharpens. A programme is under pressure to move from prototype to procurement, from trial to deployment, from promising concept to funded capability. Tests that are legible, bounded and easy to report rise in importance. Tests that are messy, combinatory or seasonally inconvenient become harder to schedule and harder to narrate. Name the governing mechanism and the argument clarifies: institutional selection of evidence. Explain it and the implication follows. The platform that emerges is not merely the result of engineering talent. It is the result of which questions the institution allowed engineering to ask without penalty.

That is why standards deserve a double reading. On the surface they are disciplines of comparability. They define thresholds, conditions, procedures and documentation so that claims can be examined rather than merely asserted. This is their virtue. Yet standards can also become shelters behind which organisations hide from the harder demands of the theatre. A system may satisfy an accepted test regime and still remain poorly matched to Arctic use if the standard does not force the combinations of cold, moisture, handling, visibility, recovery stress and sustainment difficulty that the real environment will impose. Compliance then becomes a local truth with global consequences. The paperwork is accurate. The conclusion is not large enough.

A useful historical example lies in aviation and defence procurement more broadly, where decades of experience have shown that formal qualification can coexist with awkward service reality when operational assumptions drift faster than standards do, or when programmes use minimum compliance as a substitute for representative challenge. The relevance here is structural. Standards are necessary, but they are written by institutions with finite imagination, finite time and finite appetite for embarrassment. A second example can be drawn from public procurement cultures across many technical domains, where milestone-driven reporting consistently rewards evidence that is neat enough to defend upward rather than searching enough to unsettle confidence. That pattern is not unique to Arctic systems, but the Arctic punishes it more quickly because the environment is so efficient at exposing what paperwork left out.

Still, a fair limitation should be admitted. No institution can test everything, fund every desirable trial or tolerate endless qualification drift in the name of theoretical completeness. Procurement needs decisions. Standards need closure. Programmes need movement. If every uncertainty had to be extinguished before acquisition proceeded, nothing would ever field. That is a real constraint, and serious people should acknowledge it plainly. The alternative to perfection is not automatically corruption.

But that limitation only sharpens the deeper point. Because no system can be tested against every possible harsh interaction, the organisation's incentives become even more decisive. What gets chosen as representative? What counts as an acceptable residual uncertainty? Which failures are treated as developmental noise, and which are allowed to stop progress? Which reports travel upward intact, and which are softened into milestone language? These are governance questions before they are engineering questions. The reconciliation is not to dream of a politics-free qualification regime. It is to become honest that engineering outcomes follow what institutions reward, fear and hurry.

Reporting structures matter for exactly this reason. A maintainer's note about glove-impaired access, a tester's concern about repeated cold-soak anomalies, or a field operator's discomfort with wind-recovery behaviour can either remain a local irritation or become programme-shaping evidence depending on how reporting moves through the organisation. If bad news must survive several layers of translation before it reaches a decision-maker, it often arrives stripped of urgency and mechanism alike. If every milestone meeting quietly prefers ambiguity that sounds manageable to specificity that sounds expensive, the testing regime will begin selecting for psychological comfort rather than environmental truth. The system then becomes "ready" in the language of the chart while remaining only provisionally credible in the language of the field.

This is why readiness debates should be heard as arguments about institutional character. When one group says a platform is ready and another says it is not, they are rarely disagreeing only about thermal performance, icing response or maintenance access. They are also disagreeing about what counts as evidence, how much seasonal realism qualification deserves, whether anecdotal success can outweigh structured uncertainty, and who bears the cost if optimism turns out to have been premature. Readiness, in such cases, is not a single technical state. It is a negotiated verdict shaped by incentives. The most important engineering parameter may be the organisation's tolerance for hearing unwelcome things early.

The quiet motif here is the stamp on the folder. Once applied, it gives a seductive impression of completion. The file looks settled. The process appears to have done its work. Yet everyone with real experience knows that the stamp can mean very different things depending on what was inside the folder, what never entered it, and what was translated into safer language before anyone senior signed the page. Qualification regimes often work much the same way. The authority of the stamp is real. So is the danger of asking too little before it comes down.

That is why a thoughtful reader should approach standards and procurement claims with disciplined scepticism rather than either reverence or contempt. The right question is not whether a system met a standard. The harder question is whether the standard, the schedule and the reporting culture together forced the programme to confront the Arctic conditions that actually govern credibility. Did the institution reward repeatable evidence or merely manageable narrative? Did it fund trials likely to teach, or trials likely to pass? Did it preserve awkward field knowledge, or tidy it into optimism? Those questions turn readiness from a slogan back into an argument about responsibility.

The unresolved friction remains exactly where institutions like things to sound most settled: the moment when a platform has passed enough tests to become legible to procurement, yet not enough of the right tests to become fully legible to the theatre. At that point, is the programme certifying a machine, or certifying its own willingness to stop asking what the Arctic might still reveal?

Chapter 14

Building an Arctic Autonomy Framework

How should a serious organisation judge an Arctic autonomous system: by what it can do on its best day, or by what it still deserves to be trusted to do after cold, wind, distance, maintenance strain, poor visibility and weak communications have all had their turn? That question sits underneath every argument in this book, and it is harder than it first sounds. A platform can look impressive in isolation and still prove institutionally disappointing once sustainment burden, recovery difficulty, operator workload and degraded confidence begin drawing blood from the same margin. Capability attracts the eye. Endurance tests the truth.

That is why any useful framework for Arctic autonomy must begin by distrusting the brochure. Published endurance, payload, range and autonomy claims are not necessarily false; they are merely incomplete in the way all flattering summaries are incomplete. The High North has a disciplined way of asking the questions omitted by tidy specification sheets. How often does the machine return with less confidence than it launched with? What kind of maintenance queue follows a demanding week? What has to be stocked, cleaned, dried, warmed, verified and explained before the next sortie is honest? How much of the system's apparent competence depends on circumstances that are generous rather than representative? The harder the environment, the more these questions stop being secondary and become the real architecture of usefulness.

The pivot comes here. Arctic autonomy is often discussed as though it were a problem of advanced machines entering a difficult theatre. It is better understood as a problem of institutions deciding what counts as an acceptable relationship between machine promise and operational truth. The framework needed is not a scorecard for admiring technical sophistication. It is a way of asking whether the whole system - airframe, sensors, software, operators, maintainers, spares, shelters, control logic and governance - can keep its claims intact when the environment begins taking payment in small, cumulative instalments. The wrench on the bench matters as much as the wing in the air.

This matters now because organisations routinely overestimate what can be delegated safely when they mistake autonomy for independence. A machine that flies without continuous human control may still depend intimately on clean data, disciplined maintenance, coherent ground systems, realistic recovery options and staff who understand when to narrow ambition rather than press on. If those dependencies are not examined together, the institution will continue confusing reduced hands-on control with reduced organisational burden. The Arctic punishes that confusion without sentiment. It turns neglected sustainment into mission loss, vague control doctrine into brittle behaviour, and seasonal turnover into institutional amnesia.

It will matter later for the same reason, only more so. A platform that disappoints once can be explained away as immature. A programme that disappoints across rotations, winters and crews is revealing something deeper: that its learning loop is weaker than its rhetoric. This is why the final framework cannot merely rank machines by how well they perform in discrete trials. It must ask whether the programme surrounding the machine becomes wiser, more precise and more disciplined with exposure, or whether each season starts by forgetting what the last one paid to discover. Long-term progress in this theatre is less about chasing a perfect machine than about building an organisation that can absorb evidence without flattering itself into premature certainty.

The institutional stakes are especially sharp because readiness debates are so often framed as technical disagreements when they are really disagreements about what evidence an organisation is willing to honour. One side points to successful sorties, elegant autonomy functions and persuasive demonstrations. Another points to maintenance burden, edge-case fragility, poor recoverability and the quiet labour of keeping the system credible. Both may be speaking truth. The framework that matters is the one capable of holding those truths in the same hand. A machine that is tactically dazzling but sustainment-hungry may be less useful than a less glamorous platform whose degraded modes are honest and whose maintenance rhythm does not lie. A highly autonomous function may be admirable in principle and still badly matched to the realities of poor visibility and intermittent links. The point is not to become anti-technology. It is to become suspicious of claims that do not carry their own operating costs in full view.

That suspicion should mature into discipline rather than cynicism. The framework taking shape here is meant to give the reader a durable way of comparing systems without collapsing into either fetish for innovation or nostalgic distrust of it. One thread runs through reliability, sustainment burden, recoverability and the material questions of what the platform asks from people and infrastructure after the sortie. Another runs through control limits, degraded modes and the sterner question of what kinds of action still belong to human judgement when confidence is thinning. A final thread follows the system through time, asking whether testing, maintenance evidence and operator experience are being folded back into a learning structure strong enough to survive staff rotation, procurement churn and the organisational temptation to reset the story each season.

The everyday motif here is the ledger. A serious Arctic programme keeps one whether it is formal or not, because every sortie, every maintenance action, every deferred doubt, every configuration change and every awkward recovery is writing against the balance sheet of trust. Some systems spend that trust quickly, buying impressive days with burdens that arrive later in the workshop, in the data archive or in the next crew's uncertainty. Others spend more slowly and more honestly. The most important judgement a framework can make is not which machine looks strongest at the point of sale, but which one leaves the ledger in better order after repeated contact with the theatre.

This final movement therefore does not ask the reader to choose between engineering and governance, or between hardware and doctrine, or between autonomy and human restraint. It asks something more exacting. Can the system be judged in a way that makes all of those elements visible at once? Can reliability be treated as an ecosystem rather than as a component property? Can delegation be disciplined by explicit control limits rather than inflated by marketing language? Can progress be measured across winters rather than across presentations? Those are harder standards than most public discussions prefer. They are also closer to the only standards the Arctic has ever respected.

The unresolved question that now matters is simple to state and difficult to answer honestly: when a platform is offered as an Arctic solution, is the organisation looking at a machine, or at a whole operating bargain involving maintenance, spares, degraded modes, human authority and seasonal learning that may prove far more demanding than the machine itself first appears?

Evaluating systems by reliability and sustainment

In 1914, when Ernest Shackleton's *Endurance* entered the Weddell Sea, the expedition did not fail because courage was absent or because the map had been read carelessly. It failed because survival in polar conditions always depends on a harsher question than ambition likes to ask: what remains usable, repairable, movable and trustworthy after the environment has finished editing your plans? The ship was impressive until the ice began charging its own terms. The inventory told one story. The theatre told another.

That is the right place to begin when judging autonomous systems for Arctic use. People are naturally drawn to the visible metrics first. Endurance, range, payload, autonomy features, sensor sophistication, launch flexibility, communications reach. These are not trivial. But in the High North they are not decisive by themselves. A platform that promises long persistence yet returns with recurring maintenance burden, fragile recovery assumptions, awkward spares demand and opaque failure behaviour may be less useful than a less glamorous machine whose performance is narrower but more repeatable. The poster on the wall and the workbench log are rarely telling the same truth.

This is why reliability and sustainment should be treated not as support issues but as the real measure of operational usefulness. Show the common decision error and the problem becomes obvious. Two platforms appear side by side in a briefing. One offers better brochure performance. The other looks plainer. The comparison then drifts instinctively towards the visible capabilities of flight rather than the hidden costs of keeping either machine honest through a northern season. Name the discipline missing from that scene and the argument sharpens at once: evaluation by operating burden. Explain it and the implication follows. A serious comparison asks not only what the aircraft can do when fresh, warm and fully serviced, but what sort of maintenance queue, spare logic, recovery risk and operator doubt it leaves behind after repeated use in cold, wind, moisture and poor visibility.

Reliability, in this context, should not be reduced to a sterile question of whether the machine fails or does not fail. What matters is the pattern of failure, the recoverability of function, and the speed with which the organisation can return the system to truthful service. A platform that suffers small but frequent anomalies may be more operationally corrosive than one that fails less often but more clearly, because constant low-grade doubt consumes maintenance time, operator confidence and decision margin.

A machine that degrades gracefully, preserves logs, signals uncertainty honestly and accepts conservative modes may be more valuable than one that remains dazzling until the moment it becomes inscrutable. Reliability is therefore not simply the absence of breakdown. It is the character of trouble.

That character becomes legible only when sustainment enters the frame. A drone in Arctic service is never just an aircraft. It is also a rhythm of battery conditioning, drying, cleaning, inspection, packing, connector care, sensor protection, software state, spare access, shelter quality and maintainers' judgement. This is the sustainment ecosystem, and it produces usefulness more reliably than any isolated design flourish. If the aircraft demands rare parts, delicate repair, frequent recalibration, specialised handling or interpretive maintenance rituals that only a few experts can perform, then its effective capability is already narrower than its flying qualities suggest. The system may be technically excellent and institutionally exhausting. In the Arctic, that is often a losing bargain.

A useful historical example lies in aviation operations more broadly, where fleets have long been judged not only by their nominal performance but by dispatch reliability, maintenance man-hours, parts burden and the stubborn reality of what it takes to keep them available. The lesson worth borrowing is simple and severe: the machine that looks strongest in isolated performance trials may not be the one that produces the most truthful sorties over time. A second example comes from polar field science and remote logistics, where the most cherished equipment is often not the most advanced object in abstract engineering terms, but the one that can be cleaned, trusted, repaired, transported and re-used without turning every routine action into a minor negotiation with the environment. The Arctic teaches respect for the unshowy.

This is why failure rates should be read as ecological signals rather than as bare counts. A platform's failures matter, of course, but what matters just as much is where they cluster. Do they appear in connectors, seals and exposed interfaces after glove-handling and thermal cycling? Do they accumulate in batteries after repeated cold-soak and reconditioning? Do sensors lose credibility faster than they physically fail? Do recovery-related stresses create downstream maintenance that brochures never price in? Each pattern reveals something about the system's true bargain with the theatre. A platform whose failures are sparse but strategically disruptive may be worse than one whose faults are common but quickly contained.

The intelligent question is not only "How often does it break?" but "What kind of work does its breaking create, and for whom?"

Still, a fair limitation should be admitted. Not every operational judgement can be reduced to maintainability alone. There are missions where payload quality, reach, or a particular flight profile may justify higher sustainment burden. There are cases in which a more maintenance-intensive system brings enough unique capability to warrant the strain. It would be simplistic to pretend that the plainer, easier platform is always the wiser choice. Harsh environments do not abolish trade-offs. They merely make them harder to sentimentalise.

But that limitation only clarifies the standard. The right question is not whether a capability is worth some burden. The right question is whether the institution sees the burden clearly enough to call the bargain by its proper name. If a platform needs unusually specialised spares, demands careful shelter conditions, or narrows recovery choices so sharply that each sortie spends more organisational confidence than the mission appears to earn, then those costs must be carried inside the comparison rather than hidden in separate support annexes. The reconciliation is exacting but healthy: compare platforms by the whole price of truthful use, not by airborne performance alone.

Recoverability matters especially because it reveals whether a system fails in ways the organisation can absorb. A drone that suffers trouble but can still preserve data, enter a safe degraded mode, be located, recovered and returned to diagnosis remains inside the sustainment ecosystem's ability to learn. One that disappears into uncertain link loss, ambiguous navigation failure or irrecoverable damage imposes a different kind of cost. It does not merely reduce availability. It corrodes explanation. A platform should therefore be judged partly by what happens after it ceases to be fully successful. Does failure remain governable? Does the machine become an object of maintenance and evidence, or merely of hope and blame? That is not a secondary question. It often determines whether the organisation gets stronger or merely more defensive.

The quiet motif here is the ledger. Every sortie writes in it. A clean recovery, an awkward inspection, a missing spare, a connector that no longer inspires confidence, a battery that took longer to warm, a software quirk that only appears after cold transport, a shelter too cramped for honest maintenance. Some systems fill the ledger with hidden debt while still appearing impressive in motion. Others look less theatrical in flight and leave the account in better order at the end of the week. The wiser framework notices the ledger before it notices the brochure.

This is also why comparisons across platforms should be framed in questions rather than slogans. What must be stocked forward for this machine to remain credible? How much of its maintenance depends on scarce expertise? Which faults produce manageable work, and which produce ambiguity? How often does recovery preserve the aircraft in a state that still teaches? What happens to operator confidence after a difficult cycle of launch, flight, recovery and reservice? How much of the system's advertised autonomy is actually sustained by fragile ground truth, delicate upkeep or procedural heroics? Such questions do not eliminate disagreement. They improve it. They move judgement away from admiration and towards evidence.

The deeper point is that operational usefulness is not an intrinsic property of the platform. It is an emergent property of the sustainment ecosystem surrounding it. That ecosystem includes tools, shelters, logs, training, spares, maintenance intervals, software discipline, recovery doctrine and the institutional willingness to tell the truth when the burden of keeping the machine honest begins to outweigh the grace of watching it fly. A platform without that ecosystem may still perform beautifully in moments. It will not mature into a dependable Arctic capability.

The unresolved question, then, is not which machine promises the most from the podium or the brochure. The sharper question is which one, after repeated cold sorties and ordinary northern indignities, still leaves behind a maintainable, recoverable and explainable system that the organisation can afford to trust again tomorrow.

Control limits, degraded modes, and safe delegation

On 1 June 2009, Air France Flight 447 entered a sequence in which automation, human judgement and degraded information no longer fitted together cleanly. The event belonged to crewed aviation over the Atlantic, not to Arctic drones over snow and sea ice. Yet its lesson travels with unusual force. A system can remain highly capable and still become unsafe when nobody has decided clearly enough, in advance, what the machine should keep doing, what it should stop doing, and what must pass back into unmistakably human judgement once confidence begins to thin. Capability was not the problem alone. Boundaries were.

That is the right place to begin when asking how autonomous is enough in the High North. The phrase invites a technical answer, as though one could assign a level and be done with the matter. The Arctic refuses that comfort. Visibility collapses gradually rather than theatrically. Links degrade unevenly rather than cleanly breaking. Navigation confidence narrows in ways that are difficult to describe in a single number. Recovery options diminish without necessarily disappearing. A machine may still be flying well while the conditions that once justified a certain degree of delegation have quietly gone stale. The question is therefore not how much autonomy sounds impressive. It is how much delegation remains honest as certainty drains away.

This is why control limits matter more than autonomy labels. Show the ordinary operational scene and the point sharpens at once. A platform is navigating in poor contrast with intermittent communications, modest wind disturbance and a sensor picture that is still useful but less trustworthy than it was ten minutes earlier. It can continue the route, hold, simplify the mission, return, divert, or remain steady while awaiting human instruction that may itself be delayed or based on a partial picture. Name the real problem and the argument clarifies: governance under degraded confidence. Explain it and the implication follows. The machine should not be judged mainly by whether it *can* continue, but by whether the organisation has declared clearly enough what it *may* continue doing when the reasons for earlier confidence have weakened.

That distinction separates machine execution from human governance. In strong operational design, the platform may still execute many things more quickly and more steadily than a remote human ever could. It can hold attitude, stabilise against gusts, maintain carefully bounded routes, manage energy within known logic, preserve records, enter safer degraded states and keep the vehicle inside a shrinking but still legible envelope. These are excellent reasons to automate. They are not reasons to blur authority. The human should continue governing the matters that depend on judgement about legitimacy, proportionality, tolerable uncertainty and mission worth: whether the objective still deserves pursuit, whether the machine's confidence remains good enough to authorise another step, whether recovery risk has become disproportionate, and whether the information available to the platform is still rich enough for its local competence to remain morally and operationally acceptable.

A useful historical example lies in modern civil aviation's long struggle with automation surprises, where inquiries repeatedly found that aircraft and crews became most vulnerable not when automation existed, but when mode logic, confidence boundaries and human expectations were poorly aligned. The lesson worth borrowing is not a vague warning against advanced systems. It is more exact. A highly capable machine becomes dangerous when its local actions continue to look fluent after the conditions justifying those actions have changed faster than human understanding or governance can keep up. Arctic autonomous systems inherit that danger in a more geographically isolated and logistically severe form. Their challenge is not merely to execute. It is to become smaller, simpler and more conservative at the right moment without mistaking that contraction for failure.

A second example can be drawn from remote robotics and planetary exploration, where meaningful autonomy has often consisted not in unlimited machine discretion but in carefully bounded execution under sparse communications and delayed human involvement. The key lesson is again one of disciplined sufficiency. A system does not need broad freedom to be valuable. It needs the right freedom for the conditions it can still interpret honestly. That principle becomes even more important in the Arctic, where poor visibility and intermittent links tempt organisations to over-delegate out of convenience. If the human cannot conveniently supervise every fine movement, it may seem attractive to authorise wider machine initiative. Yet convenience is a poor architect of control. Safe delegation is built from explicit limits, not from the hope that necessity will somehow make blurred authority wise.

This is where degraded modes become conceptually central. They should not be imagined as humiliating half-measures or as mere emergency leftovers after the real system has failed. A degraded mode is often the most mature expression of autonomy precisely because it tells the truth about what the machine no longer knows well enough to do. Under poor visibility, that may mean narrowing route choices, refusing aggressive recovery attempts, preserving wider margins, reducing mission ambition or prioritising stable containment over interpretive finesse. Under intermittent links, it may mean acting in ways that remain legible and recoverable rather than trying to preserve a thin illusion of full mission continuity. A good degraded mode is not a technical apology. It is a disciplined answer to the question of what usefulness still survives once precision has become too expensive.

Still, a fair limitation must be admitted. Overly restrictive control limits can make a system timid, burdensome and operationally disappointing. If too much remains human-governed, then intermittent links, latency and operator workload may cripple a platform that could otherwise have contributed meaningfully. A machine forced to seek permission for every adaptive act in a theatre hostile to continuous supervision may not be safer. It may simply be paralysed. Likewise, degraded modes can be designed so conservatively that they preserve the aircraft while surrendering too much mission value too early. This is a genuine risk, and serious organisations must acknowledge it rather than treating every call for tighter governance as proof of wisdom.

But that limitation only clarifies the standard. The right answer is neither maximal autonomy nor ceremonial human control. The right answer is to ask which functions remain computationally strong under degraded sensing and link conditions, and which decisions become more political, ethical or context-dependent precisely as the environment becomes harsher. Stable low-level execution often remains machine-friendly longer than mission interpretation does. Precise actuation may still be trustworthy after broad situational meaning has begun to erode. The reconciliation is therefore stern but usable: delegate what the machine can still do with reliable local competence, retain human governance over what depends on disputed meaning, acceptable loss, legal or organisational accountability, and the judgement that a mission objective still deserves the risks required to pursue it.

This has practical consequences for how one should think about poor visibility. Whiteout and near-whiteout do not merely degrade perception; they redistribute authority. They make it harder for humans to cross-check the machine's picture and harder for machines to justify wide initiative on the basis of thin sensor confidence. The temptation in such conditions is either to cling sentimentally to human override or to retreat lazily into full machine execution because the human cannot see enough to help. Both instincts are crude. A more disciplined approach asks what the human can still authorise sensibly, what the machine can still execute safely, and which actions now require the system to become intentionally less ambitious than the mission plan once assumed. Safe delegation is the art of making that narrowing explicit.

Intermittent links impose a similar discipline. When communications begin to fray, the question is not whether the operator still technically sits somewhere in the control loop. The harder question is whether the loop remains meaningful enough to support the kind of governance the organisation claims to be exercising. A delayed human command arriving into changed local conditions may be worse than a bounded machine behaviour already designed for that sort of uncertainty. Yet the existence of bounded machine behaviour does not authorise unlimited independence. It authorises only the behaviours the organisation has already examined, tested and accepted under exactly these degraded circumstances. Safe delegation therefore depends on prior thought more than on live cleverness. The machine cannot invent legitimate authority in the weather.

The quiet motif here is the handrail. On a clear staircase one scarcely notices it. Under dim light, carrying weight, it becomes the line that keeps movement disciplined rather than theatrical. Control limits serve the same function. They are not there to make the system look bold. They are there so that, when visibility weakens and the link stutters and the aircraft still appears just capable enough to tempt one more assertion of confidence, someone - human or machine - has already been given a line it must not cross. A system without such handrails may seem freer. It is often merely less honest about where it begins to need luck.

This also changes how "autonomous recovery" or "autonomous mission execution" should be judged. The serious question is never only whether the vehicle can complete the route, approach or task without fresh input. It is whether the autonomy knows when to stop being so proud. Does it preserve evidence when confidence narrows? Does it select safer, duller actions when interpretive certainty drops? Does it avoid trapping the human into late, brittle intervention by making its own confidence boundaries legible early enough? Does the governing organisation know which recovery attempts, holds or diversions remain authorised under degraded information, and which should be ruled out before they become emotionally tempting? These are the marks of mature delegation. They indicate a system that understands not only motion but licence.

The deeper point is that "how autonomous is enough" is the wrong question if heard as a contest for the highest possible number. Enough for what, under which confidence conditions, with what sustainment burden, with what human oversight, and at what cost to recoverability and explanation? The Arctic pushes every one of those qualifiers to the front. It exposes the difference between autonomy as an engineering achievement and autonomy as an operating bargain. The first can be dazzling. The second must be governable.

The unresolved question, then, is not whether the machine can be made to do more when the world outside becomes harder to read and the human voice reaches it less reliably. The sharper question is whether the organisation has been disciplined enough to decide, before that moment arrives, which acts still belong to machine competence and which must remain inside human judgement precisely because uncertainty has begun to make every apparently capable action more dangerous than it first appears.

Designing for learning across seasons

In 1912, when Roald Amundsen's polar success was being recognised for what it was, the decisive achievement did not lie in one brilliant dash south alone. It lay in the way previous journeys, failed assumptions, material trials and hard adjustments had already been folded into a system of practice tough enough to survive the next season's questions. Polar competence was not a single performance. It was accumulated correction. Winter kept the notes.

That is the right place to begin any serious closing argument about Arctic autonomy. Too many programmes still behave as though a successful deployment proves maturity, as though one season of tolerable performance, one strong demonstration or one cluster of encouraging sorties can settle the matter of readiness. The Arctic does not work that way. It changes the machine, the operators, the maintenance burden, the logistics picture and the institution's own confidence through repeated contact. The platform that looked convincing at the start of a season may reveal its real character only after cold soak, handling wear, spares friction, operator turnover, software patches, weak recovery days and the quiet accumulation of small compromises. Progress is seasonal.

This is why Arctic autonomy should be framed not as a discrete fielding event but as an iterative programme. Show the ordinary institutional mistake and the point sharpens quickly. A platform completes a difficult deployment, the organisation harvests a few lessons, perhaps changes a checklist, perhaps replaces a vulnerable component, perhaps updates a confidence statement, and then the rotation changes. Different crews arrive. Some maintainers leave. A software baseline moves. Procurement pressure asks for momentum rather than reflection. By the following winter, much of what was learned remains technically available and operationally diluted. Name the real problem and the argument clarifies: seasonal amnesia. Explain it and the implication follows. A system that does not carry its own learning across people, winters and configuration changes is not truly improving. It is repeatedly reintroducing itself to the same environment.

That is why testing, maintenance data and operator feedback have to be treated as parts of one learning system rather than as separate administrative categories. A field trial reveals something under controlled pressure. A maintenance log reveals something different under repeated use. Operator narratives reveal what the formal data often misses: hesitation, workload, confidence loss, awkward handling, unclear transitions, the point at which a technically acceptable behaviour stopped feeling honest. Each source is incomplete alone. Together they begin to describe the machine as it really exists in the theatre. The Arctic is unusually good at punishing organisations that let these streams remain separate. It creates failures not only in hardware or software but in translation between what was tested, what was repaired and what was felt in live use.

A useful historical example lies in aviation safety culture, where the mature lesson was never simply to collect more data after each incident, but to connect flight records, maintenance history, crew reporting and configuration knowledge tightly enough that tomorrow's aircraft became measurably wiser than yesterday's. The relevance here is direct. Arctic autonomous systems will not improve by flight testing alone, nor by maintenance statistics alone, nor by operator anecdote alone. They improve when those forms of evidence are forced into the same institutional conversation before memories fade, people rotate and software versions quietly move on. A second example comes from polar scientific operations more broadly, where the strongest programmes long ago learned that a season's findings are only as durable as the routines that preserve them through staff turnover, equipment modification and the impatience of the next field plan. The weather returns with old questions disguised as new tasks.

This is why rotation and turnover matter so much. Organisations often speak as though knowledge naturally lives in reports, databases and lessons-learned documents. Sometimes it does. More often it lives for a worrying length of time in the heads and habits of the people who froze their hands on the connectors, mistrusted the recovery cue, noticed the recurring battery lag or learned which sensor warning was nuisance and which one was the start of a harder afternoon. When those people leave or shift role, the programme discovers whether it has been learning institutionally or only personally. The distinction is severe. A good Arctic learning system captures the judgement embedded in experience without flattening it into dead paperwork. It turns tacit craft into usable memory before the season changes hands.

Still, a fair limitation should be admitted. Not every seasonal difference points to progress or regression. Winter conditions vary. Missions vary. Teams vary. Some lessons are too local to generalise cleanly, and some apparent patterns are little more than the theatre's changing mood. There is also a danger that organisations, eager to prove they are learning, begin changing too much between seasons: too many modifications, too many new procedures, too many moving baselines. Then comparison itself becomes harder, and the programme mistakes restless adjustment for disciplined improvement. The answer is not to freeze the system in place. It is to treat change as something that must itself remain legible.

That legibility depends on the way evidence is carried forward. A serious programme should be able to say, at the start of a new season, not only what changed but why, on the basis of which maintenance patterns, which test results, which operator concerns and which recoverability lessons. It should also know what *did not* change, so that new observations can be meaningfully compared with old ones. This is where the long-view model begins to matter. A machine is not being judged only against the previous sortie or the previous trial, but against a growing seasonal record of how it behaves under real Arctic burdens. The organisation's task is not to chase novelty. It is to improve while preserving enough continuity that improvement can still be recognised as such.

This gives operator feedback an especially serious status. In many technical cultures, subjective reports are treated as lesser evidence than logs, traces or engineering measurements. That hierarchy is understandable and often useful. It becomes dangerous when it leads programmes to ignore the human perception of narrowing honesty. Operators often know before the formal data does when a system's behaviour is becoming harder to trust, more awkward to recover, more fatiguing to supervise or less interpretable under poor visibility and intermittent links. Such observations are not final verdicts. They are early warnings. A learning system that cannot absorb them without embarrassment will continue privileging the cleaner signal over the truer one. In the Arctic, that is rarely a sustainable habit.

The quiet motif here is the notebook in a coat pocket. It gathers smudged observations, times, small failures, improvised fixes, remarks that seemed too minor for a headline report and too important to trust to memory. Left alone, it is personal and fragile. Folded into a disciplined programme, it becomes part of the institution's winter memory. Arctic autonomy needs something like that notebook at every scale: in testing, in maintenance, in operations and in governance. Not because anecdotes should outrank evidence, but because evidence is often born in the first place from someone deciding that a small repeated discomfort deserves a permanent line on the page.

This is also why wishful thinking has to be treated as a seasonal hazard rather than as a flaw of personality. Every programme wants to believe that each new winter begins from a higher platform of maturity than the last. Sometimes it does. Sometimes what has really accumulated is only smoother language about old limits. The difference lies in whether the organisation can point to linked evidence across seasons: testing that changed because maintenance data exposed a pattern, maintenance decisions that changed because operators kept reporting the same burden, doctrine that changed because degraded modes proved more honest than ambitious autonomy, configuration changes that remained traceable enough for the next rotation to judge their effect. Progress without those links is often just mood with a funding line.

The deeper point is that Arctic autonomy should be built as a continuing institutional argument with the environment. The theatre keeps asking the same families of question - about cold, wind, maintenance, perception, control limits, recoverability, ambiguity and trust - but it asks them through slightly different weather, different crews and different combinations of stress. A programme that expects final answers will end up disappointed or self-deceived. A programme that expects recurring questions can become much stronger. Its goal is not to close the argument but to become better at answering it season after season, with less romance, less surprise and more cumulative precision.

The unresolved question, then, is not whether a given deployment succeeded well enough to justify the next one. The sharper question is whether the organisation has built a learning system strong enough that, when crews rotate, components change, software drifts and the next winter arrives with familiar problems in slightly altered form, the programme still knows more than it knew before - or whether it is about to pay again for lessons it once believed it had already bought.

About the Author

Jonas Merakai is a nonfiction writer who approaches defence technology as a question of institutions, engineering discipline, and the limits of control under uncertainty. His work is guided by a simple premise: complex systems only earn trust when their failure modes are understood, rehearsed, and governed with clarity. Rather than treating autonomy as a destination, he is interested in autonomy as a relationship between people, machines, and the environments that expose their weakest assumptions.

Merakai writes in an editorial, research-forward style that aims to be useful to readers who need more than headlines but less than proprietary technical documentation. He is drawn to subjects where strategy becomes tangible in material choices: batteries and thermal margins, connectors and corrosion, training syllabi and maintenance logs. In that sense, his attention often returns to the long history of polar operations, where exploration narratives, Cold War basing, and contemporary surveillance all share a common theme: distance punishes improvisation, and weather is an actor with agency.

As an author, he favours careful distinctions over dramatic certainty. He aims to map the boundary between what is technically plausible and what is operationally responsible, especially in settings where communications are fragile and recovery is slow. Drones in Whiteout reflects that commitment by treating reliability, sustainment, and human oversight not as supporting details, but as the foundations on which any credible concept of Arctic autonomy must rest.

www.ingramcontent.com/pod-product-compliance
Lightning Source LLC
Chambersburg PA
CBHW031320230926
51848CB00001B/33
9789377940744